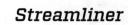

Streamliner

Streamliner

Raymond Loewy
and Image-making
in the Age of American
Industrial Design

JOHN WALL

Johns Hopkins University Press
Baltimore

© 2018 Johns Hopkins University Press
All rights reserved. Published 2018
Printed in the United States of America on acid-free paper
9 8 7 6 5 4 3 2 1

Johns Hopkins University Press
2715 North Charles Street
Baltimore, Maryland 21218-4363
www.press.jhu.edu

Library of Congress Cataloging-in-Publication Data
Names: Wall, John, 1957– author.
Title: Streamliner : Raymond Loewy and image-making in the age of American
    industrial design / John Wall.
Description: Baltimore : Johns Hopkins University Press, 2018. | Includes
    bibliographical references and index.
Identifiers: LCCN 2017044506| ISBN 9781421425740 (hardcover : alk. paper) |
    ISBN 9781421425757 (electronic) | ISBN 1421425742 (hardcover : alk. paper) |
    ISBN 1421425750 (electronic)
Subjects: LCSH: Loewy, Raymond, 1893–1986. | Industrial designers—United
    States—Biography. | Branding (Marketing)—United States—History—20th
    century.
Classification: LCC TS140.L63 W35 2018 | DDC 745.2092 [B] —dc23
    LC record available at https://lccn.loc.gov/2017044506

A catalog record for this book is available from the British Library.

Frontispiece: Raymond Loewy™/® by CMG Worldwide, Inc. / www.RaymondLoewy.com

*Special discounts are available for bulk purchases of this book. For more information,
please contact Special Sales at 410-516-6936 or specialsales@press.jhu.edu.*

Johns Hopkins University Press uses environmentally friendly book materials,
including recycled text paper that is composed of at least 30 percent post-consumer
waste, whenever possible.

For Sharon and Colleen
A better family could not be designed

Contents

Acknowledgments

This book has its roots in job interview paranoia. About five years after leaving daily journalism, I was working as a writer at Penn State's College of Agricultural Sciences when I started applying for other positions at the university. It seemed as though every person who beat me out had a master's degree. Within the academy, a graduate degree is a golden ticket.

In the back of my mind, I had always thought the story of Raymond Loewy would make a worthy book project. Armed with that idea and little else, I cold-called Stanley Weintraub, then Penn State's Evan Pugh Professor of Arts and Humanities, now happily retired in eastern Pennsylvania, and asked if I could meet with him about the master's program.

My idea was to use the master's program as an excuse to write a biography of Raymond Loewy. Stan's response was something like, "Do you want to spend a lot of time in the classroom discussing theory and writing?" My answer was, "Not really." He replied, "Forget the master's degree—just write the book."

Stan then took about an hour to lay out the roadmap to getting a book published, and I followed it to the letter. Write a magazine article that can be a sample chapter. Use the magazine article to interest presses. Pitch projects to presses with an interest in your subject, either geographical or topical. One meeting, one hour, one book (plus about fourteen years of work). My eternal thanks, Stan.

I could not have started the epic journey without the help and support of my wife, Sharon, and my daughter, Colleen. They both put up with about a decade of distracted conversations in our living room as I wrote the book from 7:00 to 9:00 p.m. each weekday. I learned early not to share every fact from my research that I found fascinating. I also learned to be present in the conversations about school, jobs, and other topics. I may have answered the question a full half hour after they asked it, but I was listening.

I would like to thank my editor at Johns Hopkins University Press, Elizabeth Demers, who rescued this project at its lowest point and brought it back to viability. Through her adept edits and thoughtful commentary, my rough manuscript became a viable book. I'd also like to acknowledge Chuck Van Hof, a retired acquisitions editor for the

University of Notre Dame Press, who supported me through all of the original writing and editing with a decade of advice and encouragement. I would like to acknowledge as well Harv Humphrey, the director of the press, who wrote a letter of recommendation when Notre Dame ultimately decided they could not publish the book. In addition, I would like to thank editorial assistants Meagan Szekely (now at USNI Press) and Lauren Straley for their help with files, invoices, and numerous other elements of publication. My eternal gratitude also goes to Johns Hopkins managing editor Juliana McCarthy, publicity manager Gene Taft, copy editor Ashleigh McKown, and many other members of the marketing and editorial staff.

I am grateful to those at Juniata College who offered words of encouragement at opportune times. Specific thanks must be paid to John Hille, retired enrollment executive vice president, who immediately granted my request for a laptop so that I could work on the book at home (this was in a long-ago dark past when laptops were relatively rare). Others who offered support include Jim Tuten, a historian of many interests; Lynn Jones, the guru of interlibrary loans; Evelyn Pembrooke, who helped me practice Loewy presentations; Jim Donaldson, a fellow car nut; Mike Keating; and Jim Watt. At Penn State, I would like to thank the university librarians who pulled countless books from the stacks for me and helped with research, and Evelyn Buckelew, my editor at Penn State College of Agricultural Sciences, who allowed me the resources to research this book.

Much-needed support for and insight into this daunting project came from Bob Mitchell, a fellow journalist and historian; Andrew Wall; Mary Marucco; J. D. Cavrich; Richard Cordray; Peggy Cordray; Carole Hwang of the Raymond Loewy Archives; research librarian Lynsey Sczechowicz at the Hagley Museum and Library; and Andrew Beckman, archivist at the Studebaker National Museum.

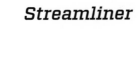

Streamliner

Introduction

In 1968, Raymond Loewy, whose instinct for the publicity photo had been unerring in the past, decided to stage an event for a meeting with the National Aeronautics and Space Administration for Loewy's Skylab commission. Heading for a client meeting at Cape Canaveral, Loewy, riding in a chauffeured Lincoln and trailed by a driver in the designer's personal Avanti, ordered the chauffeur to stop. He then managed, at age 75, to slip into an astronaut's silvery spacesuit. The aging designer, his white hair complementing the metallic gleam of the spacewear, took the Avanti and drove the sports car to the main gate of the base. Greeted by a group of quizzical security policemen, NASA engineers, and others invited for the event, Loewy claimed to have tested the spacesuit by driving four hours in the eye-catching costume.

A look at the photo today reveals a proud and rakish Loewy, with a small smile playing at the corners of his mouth, looking directly at the photographer. For most of his decades-long career, the designer had played willing subject to an endless line of photographers. His dazzling look, most often flashed as he posed insouciantly next to a car, a train, or one of his homes, had been rewarded by subsequent publication in newspapers, magazines, even the occasional newsreel.

He must have realized at the time that the media world was changing and leaving behind such publicity stunts. Still, they had worked in the past. Why change a practice that had always had the same outcome—publicity for the designer and for his design, and yet another free advertisement for his talents.

Raymond Loewy poses in a spacesuit he wore while driving onto Cape Canaveral, Florida, as a publicity stunt to tout his company's contract with NASA in 1968. (Raymond Loewy™/® by CMG Worldwide, Inc. / www.RaymondLoewy.com)

From the moment he founded his company in the late 1920s, Loewy perhaps knew instinctively, as a character in *The Man Who Shot Liberty Valance* advised, "When the legend becomes fact, print the legend." Schooled in the uniquely American art of print media and public relations by his longtime press agent, Betty Reese, Loewy was able to burnish his legend, design by design, interview by interview, article by article, throughout fifty years of the longest-lived career in industrial design.

My introduction to the legend of Raymond Loewy came in 1988 as I was interviewing a local artist for the *Altoona Mirror*. The artist was working on a painting for a book dustcover and had a reference book open to a photo of a locomotive. Its caption, barely legible in six-point type, read, "S-1 Locomotive, built in Altoona, Pennsylvania." I decided to write a newspaper feature on the locomotive. I interviewed a dozen railroaders who had worked on the S-1 project. The result, embodied in iron and steel, was the singular vision of designer Raymond Loewy. In researching Loewy's work, I read his books *Never Leave Well Enough Alone* and *Industrial Design*, and after writing my newspaper article, I moved on to other work. Over the years that followed, I read articles about Loewy's Palm Springs home and about his work on Studebakers and for Coca Cola. The articles revealed a certain consistency. The same anecdotes were repeated, and the tales of design origins have similar structure. The claims to certain designs were often cloudy. To this day, Loewy almost always is identified as the designer of the Coke bottle, although the distinctive shape was created in 1916, well before Loewy's career began.

In 1995, I decided to put together a biography of this seminal designer. His story was spread across two of his books and a wide variety of critical essays, many of them filled with contradictory information or dichotomous critical opinions presenting Loewy either as a shallow businessman who just happened to open an industrial design firm or as an underappreciated genius.

History must place Raymond Loewy in the underappreciated genius camp. In addition to his legacy of design, Raymond Loewy was a seminal creator of what is now called a "brand" as well as the creator of a business model—with himself as the showman in the center ring. Almost from the moment he stepped onto American soil, he established himself apart from other members of his profession. Whether he was the young "Frenchy" illustrator, the struggling industrial designer politicking to find the next job, or the renowned designer with homes in three

countries, he found a persona—the witty, urbane Euro-sophisticate—
that appealed to corporate executives and to the reporters and opinion
makers in emerging media.

The Loewy persona did not spring fully formed a few weeks after his
arrival. It was built over time and refined year after year until Loewy-
the-publicity-hungry-upstart became Loewy-the-elder-statesman.
Perhaps the key to the formation of Loewy's persona is his immigrant
status. With his familial ties to France cut by the influenza pandemic
after the Great War, he was free to concentrate on fitting in within a
nation where almost every family or new arrival had reinvented themselves in some way.

Loewy proved a remarkably quick study, and by using his innate
charm and sophistication, he saw that the key to American success was
in social connections. Aided by mentors such as Lucille Buchanan, a
fashion editor who introduced Loewy to his new career as an illustrator,
and his trusted image maker Betty Reese, who successfully imprinted
"Raymond Loewy, America's designer" onto the American consciousness, Loewy sought to become an American—with the missionary zeal
of the true convert.

Although Raymond Loewy's career lasted longer than that of any of
his rivals, the paper trail of his work and life is sparse. He glossed over
his early life in France in his books, and he did not illuminate much of
his experience in World War I beyond a few colorful anecdotes. Once
he established his design business, he maintained files for his clients,
but much of the paperwork, aside from some of the designer's personal
archives and the extensive files of Reese, were lost. Essayists and critics
have little to draw on in portraying Loewy's life except the same anecdotes, recycled and retold until they have the force of fact. Loewy's practiced stories about cocktail party challenges, design wagers, and Coke
bottle attributes became his biography. There is a lack of documentation
to contradict him. Reporters and researchers looking for background
ended up using press clippings repeating the same stories through decades of press coverage.

The challenge of this book is to give Loewy his due as a visionary pioneer while tracking the careful construction of his image as a designer—
as *the* American designer—from the moment he stepped off the gangway
of the SS *France*. The stories began within a year of his arrival as an un-
employed veteran with a background in engineering. Soon, as he made

inroads into his career as a freelance illustrator, came the tales of his background as a child prodigy. The well-told story of his design of the *Ayrel* (a model airplane) and the tale of his Great War trench experience were marketed first by Loewy and then by countless journalists.

Like all great creation stories, the formation of Loewy's image had truth at its core, but his propensity for the small embellishment, the oblique habit of not mentioning prominent collaborators, started early and never really stopped. This incessant polishing and buffing went on throughout Loewy's entire life. And, inevitably, the shine began to fade, first among his colleagues. At an organizational meeting of the Society of Industrial Designers, Walter Teague addressed those who protested putting forth Loewy as an officer in the association by successfully arguing that Loewy's high profile had an umbrella effect for the other designers in the society. Later, after journalists and tastemakers turned toward praising the International Style of architecture and Bauhaus-influenced products, Loewy slowly slid, if not into obscurity, then into a less exalted place in the design universe.

This book is also the story of the work behind the legend. The work directly controlled by Loewy was not on the drawing board, where he might have painstakingly produced design, like contemporary Bob Gregorie, but rather in the selling of his services, of his taste. In the long run, Loewy's legacy is one of editorial strength. Like print editors such as Harold Ross, Harold Hays, and Tina Brown, Loewy oversaw hordes of talented artists who received unprecedented opportunities to work on many varied accounts. His employees could opt to work on nothing but product design, or cars, or retail spaces. Or they could take breaks from their specialties and take assignments in other disciplines.

The longer he lived, and he lived decades longer than most of his contemporaries, the less his legend retained its gloss. Critics focused on his less stellar traits—the credit grabs and constant publicity ploys—instead of on his business acumen and image-making. To look at his work, however, is to see the record of a superb salesman. In his autobiography, Loewy tells of the grinding sameness of making sales calls, but the book comes alive in his re-creations of pitch meetings, new design rollouts, and infighting with manufacturers. He loved being the center of attention, but he was also able to deliver the final sale.

As many of his employees describe, the policy at Loewy and Associates of presenting only final designs, all with Loewy's signature prominently featured, makes it difficult to trace the evolution of a design (who created it) as well Raymond Loewy's influence on it. The company's lack of archives from its early work with Sears, International Harvester, and the Pennsylvania Railroad makes it problematic to show his influence on the staff designers.

The central chapters in this book focus on Loewy's groundbreaking work with the Studebaker Corporation, where his European sensibility and radical ideas of making smaller, more maneuverable cars kept his client afloat—in fits and starts—and, by example, eventually changed how Detroit thought about cars. The active and opinionated automotive press during Loewy's lifetime and the work of university archivists who recorded the oral histories of car designers allow us to see more clearly how Loewy's process worked and how that translated into business success.

This story is told through the designs of three important automobiles: the 1947 Studebaker Champion, which reveals how Loewy's mismanagement of a team of rivals nearly cost him his life's dream, a major automotive account; the 1953 Studebaker Starliner, which details how its major designer, Robert Bourke, was influenced by Loewy's ideas and precepts, and how other designers and car companies reacted to its debut; and the Avanti, Studebaker's doomed "halo" car that saw the most direct influence from Loewy as he personally oversaw a team of designers in a rundown ranch house in Southern California. By examining these designs—as well as Loewy's quirky "mashup" auto redesigns of Jaguars, BMWs, and Lancias—it becomes easier to see Loewy's editorial eye at work. At the same time, by tracing the design influences of these three models, the legacy of Loewy's contributions to design history comes into focus. These three automotive designs changed the car industry subtly and not immediately, but the Big Three carmakers evolved nonetheless. After Loewy took over the Studebaker account, manufacturers slowly and reluctantly recognized that the balanced equilibrium of the 1947 Champion, the lightweight and sporty drivability of the 1953 Starliner, and the space-age individuality of the Avanti were harbingers of a changing consumer market. Ultimately, Loewy could not convince Studebaker executives of his vision. Studebaker Corporation was long dead by the time the American car industry came around to Loewy's ideas. His tenets of "weight is the enemy" and high-mileage compactness would prove to be the road to riches—just not for Studebaker.

The story of Studebaker is also somewhat of a mirror for the story of Loewy as designer. With humble beginnings in a foreign and new environment, followed by steady success and expansion, Loewy's long, slow business decline was not as precipitous as Studebaker's, but many of the same factors played into it, such as Loewy's stubborn refusal to bring in an inheritor to his firm, his penchant for siphoning company funds to support a lavish lifestyle, his failure to recognize a changing marketplace that did not value "great man" multispecialty designers over in-house specialists, and a lack of interest in a product that could have kept the company afloat—compact cars in Studebaker's case, interior retail design in Loewy's case.

As the later chapters of this book reveal, the last two decades of Loewy's career found the designer making decisions based on maintaining his personal lifestyle, much like Studebaker's auto executives managing for stock dividends and gutting company pension funds to keep the company viable. His focus on personal income and maintaining his financial and social image in the end lessened his reputation among colleagues, critics, and journalists. An extensive memo written by Loewy that recounts the occasionally acrimonious meetings between himself and his partner, retail designer William Snaith, as they made the decision to liquidate the company, reveals overt anger. Loewy quotes Snaith as telling him, "All you want is to get personal publicity and make money."

Loewy did sell his US company, and concentrated on his European operation in France, and he then found himself selling off homes and his personal archive to give his family a financial foundation. In doing so, Loewy sparked widening cracks in what once was a sterling reputation. The parceling out of his papers and archive appalled most of his former employees and many in the industrial design profession. The move also generated negative publicity from many media outlets. Over time, however, even in the last few years of Loewy's life, the judgmental attitudes in the media and in professional organizations toward self-promotion slowly loosened. Doctors and lawyers started to advertise. Car executives became television spokesmen. By the 1990s, Raymond Loewy's reputation for incessantly grabbing the spotlight began to seem less like rampant egotism than a roadmap for the construction of a personal brand.

Seen through modern eyes, every phase of Loewy's life seems like a calculated campaign of image-building. His choice to place personal style at the forefront of his marketing effort turned out to be a template, for better or for worse, for twenty-first-century marketing. Many of Loewy's competitors—famously personified by Henry Dreyfuss and his brown suits—created reputations by remaining in the background, but the Loewy tics and techniques that so infuriated his contemporaries and some art and design critics turned out to be the path of, if not righteousness, then effective marketing.

A series of letters from Loewy to executives and members of the board of directors of American Motors Corporation (AMC) epitomize Loewy's effective, some might say relentless, approach to soliciting business. When Robert Beverly Evans purchased 200,000 shares of the Detroit-based auto company in January 1966 and soon described the company's models as conservative, Raymond Loewy seized the opportunity to write Evans a letter offering his services. "I realize American Motors has a large and competent design staff," Loewy reasoned, but he still extolled what his firm could do to improve the model line. Loewy knew that incoming executives want to make a big first impression. "A meeting would waste no one's time," Loewy concluded.

The Loewy treatment—the social notes disguising business solicitations, the timed cover stories for major product rollouts, the memoir disguising a business manual (or vice versa)—defines contemporary media management. If he were alive today, it is easy to imagine a Loewy Twitter feed and a Loraymo Facebook page. The relentlessness of Loewy's publicity machine, inhabited by a cultured, witty, urbane, impeccably dressed, and devastatingly handsome Frenchman-turned-American, turned out to be a template to become, with apologies to Howard Stern, the king of all media.

Although he would later be criticized by his own countrymen for being "too American" when he opened his Compagnie de l'Esthétique Industrielle design firm in Paris, Loewy learned to push himself to be the center of attention. He figured out how to deliver brilliant anecdotes, funny remarks (yet sometimes cutting if referring to competitors), and detailed "process stories" for successful designs. He went beyond the caricature of the "pushy American" to establish an entirely new category. Later, the archetype would be called "media savvy" or "name designer," but Loewy created the blueprint for others to build a reputational persona. Nothing was left to chance: the perfectly turned-out clothes, always tasteful but much more stylish than the conservative business attire of most executives, and the many houses, each designed to be used to entertain clients and to appeal to specific audiences, including the New York apartment littered with art, Loewy-designed products, and tastefully chosen furniture and the home in Palm Springs, modern from its glass-dominant architecture to its indoor-outdoor pool to its plush white rug. Every car Loewy owned, whether a Lincoln or Lancia customized by his sometimes baroque eye or a Studebaker, was designed to make an impression.

That impression came full circle. After Loewy's death in 1986, his reputation was rehabilitated—once again gradually—article by article, essay by essay. The simultaneous development of a new media paradigm, one where manipulation of newspapers, magazine, and television was celebrated instead of disdained, has put Loewy's career in, if not a new light, then in a more accepting and appreciative way.

1

New Shores
Creating a Biography on the Fly

A recurring theme in American literature is reinvention, of lighting out for the territory to find a new life elsewhere. In many of these stories, the protagonist wants to break away from a stifling existence in the city, country, suburbia, or from a previous life. Raymond Loewy, born in France and an immigrant to the United States before his twenty-fifth birthday, sought to escape from a country torn by the Great War and a family ravaged by illness. As an immigrant, he was free to eliminate autobiographical details that failed to mesh with a symmetrical life story or to embellish with extra detail that might prove difficult to verify. Raymond was complicit in the invention of his own story. He was astute enough to manage his public image nearly from the moment he emigrated to the United States. He continually edited and polished his life story throughout his career as a designer and artist.

He created an amazing cocktail of chutzpah and ego as he described himself: "A young man who came to America to make a living and simply happened to do so in a profession he helped create." That he helped create the profession of industrial design is inarguable; he created his public image and maintained it assiduously in speeches, articles in the media of the day, books, and everyday conversations throughout a six-decade career. His use of anecdotes and invented scenarios were effective tools to seal deals by entertaining his audiences. At the same time, these stories put distance between himself and clients and even coworkers.

Raymond assumes the voice of the back-fence conversationalist in his writings rather than a person bent on setting down the historical record. "Clarity was not always a strong point in Loewy's memoirs," design historian Paul Jodard writes drily. *Never Leave Well Enough Alone*, which has been described as "a 100,000-word after-dinner speech," opens with Raymond's exploits as a soldier and officer in the French Army in World War I and recounts a classic iteration of the immigrant success story. With practiced timing and emphasis, the author creates an "origin story" capable of competing with any politician, business titan, or celebrity of his time.

He recounts his exploits as a captain in France's Army Corps of Engineers, a career that started in 1914. He never explicitly mentions combat, despite having earned several medals, preferring to gloss over his participation in many of the war's battles. He enlisted as a private first class and was promoted to officer status after the army recognized his talent as an engineer. Although the truth would certainly add to the dramatic arc of the story (he crawled into no-man's-land to repair communication lines), Raymond prefers to talk about his sense of style. In an effective war story (military slang for an embellished tale), repeated so often over his career that it was included in most of the profiles written about him until his death, Raymond tells of creating a sanctuary of taste and design in his trench quarters. Dubbing the shelter "Studio Rue de la Paix" ("Studio on Peace Street"), Raymond describes liberating furniture and fine carpets out of bombed-out houses. He regales the reader with tales of bringing back wallpaper designs from Paris while on leave. His description makes the battlefield trench shelter, which certainly would have been awash in mud, vermin, and the detritus of a typical World War I trench, sound worthy of inclusion in *Architectural Digest*. By the time enterprising journalists might have questioned such a story, most evidence of trench warfare had long been buried. Raymond's wartime bravery is undisputed as evidenced by his medal citations, but other accounts contradicting his version have not surfaced.

The only published record of Loewy's early life is *Never Leave Well Enough Alone*, and, as readers can see, Raymond Loewy might be a textbook example of the "unreliable narrator"—half a century before James Frey. Some facts are indisputable. Raymond was born in Paris on November 5, 1893. His father was Jewish, his mother Catholic. His father, Maximilian Loewy, was a journalist on financial and economic issues and editor of a business journal. Maximilian was an Austrian Jew who had emigrated to France at age 20 to pursue a writing career and to escape the virulent anti-Semitism of nineteenth-century Austria. Maximilian soon met and married Marie Labalme, whose family owned land in the Ardeche region. Raymond credits her with coining the family motto, "It is better to be envied than pitied."

In Paris, Raymond and his brothers—Maximilian, born in 1889, and Georges, born in 1891—saw leaps in technology, transportation, and innovation in the decades leading to World War I. France, unlike its great rival England, entered the machine age as a follower rather than a leader. Companies retooled or remade their industries to compete with Britain, whose advances in industrial processes allowed them to make goods much more cheaply than the competition. France fostered its entrance into the industrial fray by sponsoring efforts to emulate the British system. The state created exhibitions of industrial products, sponsored contests for industries, and offered prizes to inventors and innovators. Raymond and others competed in these state-sponsored contests, spurring entrepreneurial efforts in industry and business.

France was stymied in its transition to industrial self-sufficiency because, unlike Great Britain, it lacked large deposits of coal. Unable to fuel the large-scale manufacturing plants typical of the new industrial age, the French found a competitive edge in upmarket luxury textiles. These markets were open because Great Britain dominated the other textile markets, producing miles of cloth for inexpensive linens, clothing, and other products in its industrialized mills. At the beginning of the Industrial Revolution, roughly 1815 to 1860, France was still primarily an agrarian society, remaining largely rural well into the nineteenth century. It also grew into a professional and industrial nation, "the regime of veterinarians and pharmacists." In Raymond Loewy's formative years, however, France made headway in another industry besides textiles: the fledgling automobile industry, giving young Raymond the opportunity to see a variety of automotive makes, from horseless carriages to phaetons.

The Loewys valued education, sending all three of their sons into the professional classes. The rise of the French middle class was driven by advancement of education. According to historian Jeremy Popkin, many Jews rose to prominence in academia, professional careers, and journalism—including Maximilian Loewy. But several years before

Raymond was born, the specter of anti-Semitism, never far below the surface in many European countries, arose during the Dreyfus Affair, which roiled the fabric of France.

Alfred Dreyfus, a 35-year-old captain in the French artillery from Alsace-Lorraine, fled to France to escape being drafted into the German Army. In 1895, he was accused of giving a memorandum to the German embassy revealing the firing mechanism for a French cannon. Despite little evidence, he was convicted in a brutally fast trial and sentenced to incarceration on Devil's Island, the French penal colony. At the same time, many Jews, like Maximilian Loewy, were escaping Germany and Austria for better opportunities in France. The influx raised resentments and anti-Semitism in Paris and other population centers. To most of the French population, Dreyfus got what he deserved. Over time, however, a small but vocal group of opinion makers actively supported Dreyfus. Emile Zola, France's popular novelist, took up the cause, publishing *J'accuse!*, a heartfelt open-letter defense of the accused captain. A French intelligence officer, Marie-Georges Piquart, discovered that the disputed memoranda was a forgery, but the French General Staff refused to grant a pardon, and jailed Piquart for his trouble.

The railroaded Dreyfus eventually gained his release, but not before anti-Semitic riots broke out in seventy French cities in 1898 and Jews were made to feel alien in their own (or adopted) country. The scandal, which was covered by journalists throughout France and Europe, certainly could not have escaped the notice of Maximilian Loewy.

As a business journalist, Maximilian reported on the growth of individual enterprises and the collective prosperity of his adopted home. France was eventually recognized as a modern industrial center, thanks in large part to two "universal expositions." These world's fairs brought opinion makers from all over the world to Paris. Citizens of the nation talked to each other via a telephone system "functioning in 1879, in Paris, about as efficiently as it would for the next century." The next exposition, in 1899, was dominated by the Eiffel Tower. The Loewy family would have seen the magnificent structure less than a half mile from their Opera District neighborhood.

The tower was meant to mark the centennial of the French Revolution. Gustave Eiffel, an engineer born in Alsace-Lorraine, had designed metal bridges and viaducts, and proposed his plans for the tower as the world's tallest structure—"the only country with a 300-metre flagpole," he wrote. Eiffel's structural résumé included a cast iron bridge at Garabit, locks for France's doomed excursion to build the Panama Canal, and the iron skeleton for Frédéric Auguste Bartholdi's Statue of Liberty. The price tag for the tower was 15 million francs, and it was dedicated on March 31, 1889. At the ceremony, sixty people climbed the tower, with the rotund Prime Minister Pierre Tirard giving up after ascending to the first platform. The technology used to construct it was astounding. Each strut and structural member was predrilled and prefabricated to Eiffel's exacting specifications. Now the pride of nearly every French citizen, the tower was greeted by lukewarm reviews by Parisians. Poet Paul-Marie Verlaine promised never to visit the site again, and observers called the structure unflattering names, the tamest of which were "metal asparagus," "hollow candlestick," and "solitary suppository."

Raymond Loewy saw the tower as the symbol of Paris throughout his lifetime. During his early life, Paris grew into the cultural center of Europe. The Metro subway system opened in 1900, each station in the city decorated with Art Nouveau metalwork created by Hector Guimard. By the start of World War I, the two celebrated rail stations Gare de Lyons and Gare d'Orsay had opened. French politician Guillaume Chastenet called Paris "a work of art. It is a collective and complex art, it is true, but this makes it an even higher form of art." As might be imagined, this artistic city drew a population of artists and intellectuals. James Joyce, Lenin, and Leon Trotsky lived in Paris in 1903, as did Pablo Picasso. Henri Matisse and his followers, the Fauves, were producing leading-edge artworks. In 1908, Sergei Diaghilev's Ballets Russes arrived for a series of acclaimed performances, and one year before the start of the Great War, Igor Stravinsky premiered *The Rite of Spring* in the City of Lights.

French consumers and those in other European nations began to see more and more manufactured products appear on mercantile shelves. The Loewys, who employed servants, owned their home, and sent their children to top schools, had the means to partake in the expanding cornucopia of products crowding Parisian stores. Although industrial design had yet to be recognized as a profession, the rest of Europe had recognized that products could be created with both utility and desirability. At the turn of the century (through the start of the Great War), the most influential designer was Peter Behrens, who created a series of influential designs for the German company AEG (Allgemeine Elektricitäts Gesellschaft). Behrens in turn hired and trained three designers who would set design standards for the next five decades: Walter Gropius, Mies van der Rohe, and Charles-Édouard Jeanneret-Gris, also known as Le Corbusier. Gropius would go on to found and oversee the center of pre–World War II design, the Bauhaus. The curriculum of the school followed a liberal arts core and experiential mission rather than the classic art-school model of sketching and reproducing other artists' work. Many of the faculty emigrated to the United States when the Bauhaus closed in 1933. Gropius and Marcel Breuer went to Harvard, Laszlo Moholy-Nagy opened the school of Design in Chicago, which eventually became the Illinois Institute of Technology, and Josef Albers led the Yale art faculty. The influx of these design specialists would shape the fate of American product design as well as cement the "less is more" and Louis Sullivan–derived "form follows function" theories that became the dominant design doctrine for the intellectuals and critics of the twentieth century.

Raymond would have seen some of the new designs created by Behrens and others, particularly the trains and housewares of the period. The Loewys were well off, making enough money to employ nurses for the three Loewy boys as well as maids and other domestics.

In *Never Leave Well Enough Alone*, Raymond spends little time telling of his background, preferring to recount colorful tales of family minutia, such as his father's conversion to vegetarianism. The topics of religion and the pressures of casual anti-Semitism went unmentioned. When Raymond wrote his memoir, nearly forty years before his death, he preferred to downplay his religious background as well as his relatively comfortable upbringing. That decision centered purely on practicality and pragmatism. The business atmosphere of Raymond's time remained plagued by anti-Semitism, particularly in many of the manufacturing centers where he had to sell his services. It was one thing to sell midwestern executives on the design decisions of a Frenchman, but adding information about his father's Judaism certainly would have been an obstacle in some business circles.

Raymond seems to have lived an idyllic life in middle-class French society. He attended Chaptal College, an exclusive elementary school. Emulating his father, he established a weekly news magazine for neighborhood readers, *La Journal de Plombieres*. Unable to resist showing off his business acumen, Raymond writes, "circulation at its peak was ten." Presumably, the readership was the immediate family and a friend or two. The topics of the newsletter seemed to have centered on transportation and food. One issue detailed a trip to the Rhine and touted the wine and food of the region. The author was 12 years old. He also wrote about cars, boats, and trains, topics that arose in letters to his brothers and parents throughout his youth.

Engineering was also a formative talent of Raymond's. In his memoir, he wrote of winning a lentil-growing competition with an irrigation system capable of self-watering plants over the weekend. The system, which used several wicks and reservoirs of water, revealed young Raymond's ingenuity as well as a keen talent for staying ahead of the competition.

Motion, preferably accompanied by speed, fascinated the young man. His youthful writings, remembered in his adult books years later, spend pages on descriptions of train rides and automobile sightings. In *Never Leave Well Enough Alone*, he wrote ecstatically of seeing, at age 10, Brazilian air pioneer Alberto Santos-Dumont fly a prototype airplane over the polo field in Raymond's favorite Paris park, the Bois de Boulogne. Like most young boys, he loved to draw machines and filled his school notebooks with images of trains and automobiles. A photo of one of his early sketchbooks reveals a thick journal crammed with notes, drawings, and calculations occupying every square inch of paper. Second only to his interest in transportation was his interest in money. His first job was reading the news to his father—a career aimed at earning cash for a bicycle. During summer vacations, he and brothers Maximillian and Georges would stay with their mother in Trouville on the Normandy coast. His clearest memory of the experience was auto-motive, watching a Mercedes deliver the *New York Herald* every morn-ing. "My greatest thrill was to feel with my hand the honeycomb texture of the radiator, still hot and regurgitating; clogged with squashed but-terflies, bugs and sometimes tiny little birds, holocausts to speed." The look and feel of machines moving through space captured Raymond's imagination early on. On trips to Nice in 1905 with his mother, Raymond would beg her to let him see the roundhouse of the PLM (Paris-Lyon-Méditerranée) Railroad. He recalled the "graceful" locomotives, saying, "They effectively influenced me the rest of my life."

By his teenage years, Raymond became enamored of yet another mode of transport, the flying machine. He begged his parents to let him go to the Bois de Boulogne to fly model airplanes. He preferred the small airplanes designed by Hubert Lolham called "Antoinettes." He joined

A young Raymond Loewy at the wheel of a soapbox car in France. (Library of Congress, Prints and Photographs Division, Visual Materials from the Raymond Loewy Papers [Reproduction number LC-USZ62-98202])

a flying club to test out their designs and maneuvers. His new hobby would yield his first inspiration for a design that others could execute. *New York Herald* publisher James Gordon Bennett sponsored an international airplane design contest. Raymond entered with a design for a model airplane called *Ayrel.* Foreshadowing his penchant for self-promotion, *Ayrel* comes from the French phonetic pronunciation of the initials R.L. The design won top prize, the 1908 Bennett Cup. The schoolchildren in Raymond's club asked him to make similar models for them, and, overwhelmed by the popularity of his design, he took out a patent and registered *Ayrel* as a trademark. He was 15.

Unlike most teenagers, prone to flitting from one enthusiasm to another, Raymond did not let his attention flag from his model project. Instead of selling the design to a manufacturer, the young entrepreneur decided to start his own company. Advised by his father, who covered a variety of businesses on the financial beat, he formed the Ayrel Corporation and immediately hired two mechanics and a salesman. An instinctive aptitude for expanding an opportunity to fill a need would serve Raymond well time and time again as his career developed. He also developed a near-fearless belief in the certainty of his ideas.

It was at this point, still in his teenage years, that Raymond developed the seminal talent for inspiring others with his ideas. He always described himself as a shy boy, but the *Ayrel* business required someone to market the model airplane. He agreed to travel for a series of lectures sponsored by the National Aeronautical League, a hobbyist organization that no doubt would be fascinated hearing the business ideas of a teenage aircraft designer. He took the stage to tentatively demonstrate the features of the plane, make a short speech about the company, and take questions from the audience. With each lecture, the young man gained confidence, although he repeatedly practiced his presentations before going onstage. He said of the trial-by-fire experience, "It helped me overcome a great shyness that had made my life miserable until then."

Young Raymond had a relatively short career as an entrepreneur. Both his parents were alarmed at how much time he was devoting to the venture and asked him to give it up. He sold the business to the salesman he had hired. Raymond did not bank his gains or hand them over to his parents. Instead, he took the proceeds from the sale and bought materials to design a model speedboat. Calling his creation the *Ayrel II*, Loewy promptly entered the design in a competition for model boats, the Branger Cup, held on the lake in the Bois de Boulogne. As he explored the beauty of creating workable boats and airplane, Loewy was intensely aware of the modern age emerging in Paris.

By 1908, the French capital had seen the introduction of electricity, radio, telephones, automobiles, and airplanes. Any process that could be mechanized joined the parade of new products that manufacturers created to make lives easier. Most of these items, especially compared to the hand-crafted, beautifully conceived creations of the previous generation, were designed by machinists whose primary focus was the reliable operation of the machine. Ease of use and appeal to the customer were not really considered. There is little doubt Raymond had access to most of these products, as his parents were solid members of the middle class.

For the Loewys, education was intrinsic to success. Raymond and his brothers were expected to go to school and to do well. Although his autobiography never overtly mentions it, the Loewys wanted their youngest son to follow his interest in engineering. At age 15, Raymond entered École de Lanneau, which functioned as a preparatory school for École Centrale, perhaps the most prestigious technological institute in France.

Like many young men who find an interest early in their lives, Raymond received grades that reflected the tunnel vision that often prevents prodigies from exploring areas beyond their focus. Raymond received excellent marks in higher mathematics, trigonometry, descriptive geometry, and mechanical design. But he received zeros in chemistry, physics, philosophy, literature, and languages. Raymond attributed the low marks to his "smart-aleck" persona. Certainly, the aptitude he showed in mechanical engineering and design made it seem to him that unrelated courses were a waste of time. The engineering curriculum captured his attention, and the future artist never relinquished his interest.

As Raymond's studies wound down, he readied himself for an engineering career, with electronics and telegraph technology piquing his curiosity. For a short time, he worked as a journeyman in an electrical engineering firm. But first, France's mandatory military service called. He entered the army as a private in 1914, and with the advent of war he remained in service. He saw fighting from the beginning, and despite his cavalier descriptions of stylishly outfitted trench quarters, he experienced extended periods of combat and was decorated for bravery three times. His unquenchable thirst for adventure may have come from his experiences in the war. Design historian Paul Jodard theorized, "His energy, even when misplaced, must command respect. One source for it that may have been overlooked is the fact of his surviving the First World War, despite having to serve all through it on the Western front."

By the date of Loewy's release from the army at the end of the war in 1918, the world he had known as a student had changed. His mother and father had both died during the Spanish flu pandemic, which, spread by troop movements, wiped out 20 to 40 million people worldwide. The flu, coupled with the millions of young men who died in the war, created the opportunity for many Europeans to create a life different from previous generations. Professions became more accessible to people of all classes. Raymond and his brothers had no family to return to, making it much easier to leave their home country for better opportunities. His parents had left little to inherit, so the Loewys' ties to France rapidly disappeared as each brother left the country. His brother Max, who had chosen a career in business, was the first to emigrate to the United States. He spent a short time in New York City and eventually settled in Mexico. His brother Georges, who had served as a medic in World War I, found a career in New York as a doctor at the Rockefeller Institute, specializing in the treatment of gas victims. His siblings' success and a series of descriptive letters to their younger brother inspired Raymond to come to the United States in 1919.

Thanks in part to his father's profession, Raymond was uncommonly familiar with American culture. His family read the *New York Herald* every day, and Raymond was proud that he knew the works of Edgar Allan Poe, Walt Whitman, and Mark Twain. He had one set of clothes, his exquisitely tailored army uniform, and had roughly $40 in cash, which had been sent to him by his brother. With that small stake, he boarded the liner SS *France*. Raymond Loewy was going to America.

2

Portrait of the Young
Engineer as an Artist

Raymond Loewy paced the deck of the SS *France* as he sailed toward an unknown future in the United States. Unable to take advantage of some of the more expensive features of the ocean liner, Loewy spent much of his voyage walking around the ship and resurrecting his prewar habit of sketching. He made drawings of ship fittings, deck scenes, and fellow travelers. During the voyage, the ship's captain organized a silent auction of various items donated by the passengers. Loewy contributed a sketch of a young woman strolling along the ship's deck.

The sketch was purchased by Sir Henry Armstrong, the British consul in New York. Impressed by Loewy's talent, Armstrong offered to make some introductions to employers once the young man came ashore. Armstrong knew several magazine owners, among them Condé Nast. Before meeting the diplomat, Loewy recalled in his autobiography, he had not planned a career in art or illustration. Two of his brothers had found jobs in the United States directly related to their training—Max in banking and Georges in medicine. Raymond was hoping to find a position relating to his education and military experience, and planned to apply with General Electric as an electrical engineer. The serendipitous onboard encounter with Consul Armstrong was the first in a series of fortuitous meetings, all the result of social networking, which steered the young Frenchman away from the engineering dreams of his early life. Instead he would pioneer another career, that of social engineer.

Maximilian Loewy met Raymond at the dock upon his arrival in the United States. The poised young man, dressed stiffly in his army uniform, stepped onto American soil unable to speak much English. He rented an apartment in a building on the east side of Manhattan, a four-story walkup. His brother introduced him to the neighbors. Not everything went smoothly. He kissed the hand of one neighbor and later joked that the woman never talked with him again. He wasted no time acquainting himself with the sometimes intimidating bustle of the city. He chose to look for work in illustration, bearing his letter of recommendation from Henry Armstrong and carrying a small portfolio of his drawings.

He wrote of his first impressions of New York, "The giant scale of all things. Their ruggedness, their bulk, were frightening. Subways were thundering, masses of sinister force, streetcars were monstrous and clattering hunks of rushing cast iron." Georges introduced Raymond to his colleagues at the Rockefeller Institute, and Raymond quickly assimilated into the close-knit society of researchers, many of them recent arrivals from France and other European nations. Through one of his brother's acquaintances, he met Herbert Strauss, heir to Macy's department store. Strauss offered Loewy a job dressing windows for the flagship store.

Loewy writes that he started working on a Sunday to create a test window. Strauss introduced Loewy to an employee who was to supervise the young Frenchman. Loewy was less than impressed by the man, writing, "He looked like a magnified version of a specimen I had seen preserved in an alcohol jar in some pathological lab." The situation did not improve when Strauss informed Loewy's supervisor that the new hire was not to be a "clock puncher." "He looked as if he wished I were the clock," Loewy wrote. His description of the job in his autobiography is the first design assignment that receives a detailed overview.

"At the time, the technique was to [bring in] a truckload of stuff . . . The result looked somewhat like the drawing room in the mansion of the eccentric Collyer brothers, who never went out in fifteen years and were found dead under a piano." Determined to break the mold for his version of a display window, Loewy dressed a mannequin in black, spread a mink stole at its feet, and haphazardly scattered a few accessories on the floor. There was no lighting, save for a single spotlight illuminating the mannequin. He wrote of it, "It was dramatic, simple and potent—it sang." When he returned to work the following day, he found his creation had hit a sour note with the Macy's staff. "They were talking in hushed tones, as if the founder's daughter had been found raped in the window."

Sensing his imminent firing, the perceptive young man beat the company to the punch by quitting outright. After his brief debacle at Macy's, he decided then and there never to work for anybody but himself. He said nearly sixty years later about the decision to go into industrial design, "I imagined that a time would come when I could combine an aesthetic sensibility with my professional background in engineering."

Loewy's one-day design career was unfortunate because department stores were indicators that the post–World War I mass market was beginning to grow exponentially. The American way of life would change to a consumer culture within two decades, with self-motivated families using more labor-saving products. The methods by which people bought these machines also evolved at a rapid pace. By the end of the 1920s, individual storeowners were being bought out, small chains of businesses merged to become larger chains, and all chain stores were marketed in the same way. New product niches appeared seemingly overnight. Perfumes and cosmetics became the tenth largest industry in the country in the 1920s, recording $1 billion in sales. Purchases of clocks and watches increased from 34 million to 82 million from 1920 to 1930. Previously unknown products such as refrigerators, toasters, vacuum cleaners, fans, stoves, and dishwashers were the top-selling products of the era.

Before television, upwardly mobile workers looking for goods sought them out in store window displays, usually while strolling along downtown storefronts. As new immigrant Raymond Loewy walked in Manhattan looking for his own opportunity, he realized that there was room for yet another fledgling artist in the city. In a supremely self-confident move that would mirror his methods for growing his businesses, Loewy used the letter of recommendation from Consul Armstrong to make a series of appointments with department stores, magazine publishers, and other businesses that might hire a sketch artist. He met with Philadelphia department store magnate Rodman Wanamaker and was hired to make several sketches for a print ad. He returned with the drawings, and the ad department immediately accepted them. Loewy left the building with the promise of more work. A meeting with publisher Condé Nast yielded an assignment that Loewy performed in the company's office, and he left with a check. He was in business. "I always abhorred the role of being a spectator," he wrote.

There were plenty of businesses to solicit as well. Retail chains emerged in countless categories—groceries, drugstores, hardware, and dry goods—which subsequently morphed into subcategory chains such as candy stores, hotels, and restaurants. By 1929, Marion, Ohio, a city that epitomized "small-town" America, featured a main street dotted with national retailers: two Kresge's dime stores, two Kroger grocery stores, three chain clothing stores, one Woolworth's, two chain shoe stores, one Montgomery Ward, and one J. C. Penney store.

As an independent artist, mainly for print ads, Loewy worked for Butterick, the White Star Line, and Pierce-Arrow Motor Car Company, and he designed several costumes for variety showman Florenz Ziegfeld from 1919 through much of the 1920s. For the first several years of his career, Loewy signed his work "Raimon," perhaps because the visual élan of the signature looked more continental and sophisticated. His experience with creating print ads proved challenging enough that he decided to focus his efforts on fashion illustration. By building his client list one assignment at a time, Loewy was able to make a living and learn more about American culture. On October 1, 1919, he wrote to his brother Max, "There is no need to worry; what I'm doing is very well thought of here and when my friends introduce me to someone I'm 'the French artist' and I can tell you that sounds pretty chic, it makes a good impression right from the start. It's much more original than engineer in a country where any Tom, Dick or Harry calls himself an engineer so to speak."

That same month he wrote to Georges: "You say one has to have personality, well I think I have it. I served 50 months at the front, I arrived in America, one month later I sold some sketches I had whipped off in an hour for 75 dollars apiece to people I didn't know, without being able to speak English, I've done designs for the trendiest magazine in the world, gowns for Ziegfeld, autos for Pierce-Arrow, and socks for Onyx Hosiery: call that a nobody?" When he asked Georges for a loan upon arriving in New York, the money went to purchase a collared shirt so he could be properly dressed if he found the opportunity to mingle with any potential "society" clients.

In 1924, he met with Henry Sell, editor of *Harper's Bazaar*, and landed his most lucrative account to date. He worked closely with Lucille Buchanan, fashion editor for the magazine. She acted as Loewy's mentor, taking him to parties, dinners, and weddings and making introductions.

At these various events, Loewy sketched countless images of "society" men and women. Buchanan, known to her friends as "Tookie," gave her young protégé invaluable insights into how Americans perceived style. Perhaps more importantly, their relationship, which Loewy pointedly described as platonic, provided much-needed insight into American humor, American tastes, and how American saw themselves. In general, most American businessmen had a self-image of visionary modernity and wanted the public to make that same connection. The connection came through the efforts of artists designing ads, packaging, and other marketing methods. "The French artist" benefitted immensely from the avid interest in the French Art Deco movement. After the 1925 Paris Art Deco exhibition, he was described in the press as one of the best Art Deco practitioners in New York. "Financially, I was successful but I was intellectually frustrated," he told the *New York Times* late in his life. "Prosperity was at its peak but America was turning out mountains of ugly, sleazy junk. I was offended my adopted country was swamping the world with so much junk."

Fashion illustration brought him into society. For Loewy, the wonderful aspect of America was that citizens could create their own class through hard work (and astute networking). Although Loewy felt he had talent as a commercial artist, he was never a master draftsman. Most of his existing work reveals an eye for geometric design rather than drawing. In his later years, most of his employees would comment on his less than stellar talent in this area.

The world of fashion illustration only got his foot in the door; it was up to Loewy to push through it. In 1927, Horace Saks met Loewy and told the young artist about his new store opening on Fifth Avenue. Loewy advised the retailer that he should create a special climate within the store, what marketers of the twenty-first century would call "a shopping experience." Loewy, in a rare instance of low-key advice, told Saks all his employees should look "neat and clean." Saks asked Loewy to design the elevator operators' uniforms, an assignment Loewy found energizing. The men's uniforms are described in *Never Leave Well Enough Alone* as dark suits lined with white pique, ascot ties, and white gloves. This single assignment yielded large dividends for Loewy's illustration career. He did much of Saks advertising illustrations for the next several years.

A benefit in becoming one of New York's top retail illustrators included exposure to the flood of consumer goods that entered American markets after the war. Stores "crammed to the ceiling with everything in the world from aspirin to roller skates" were Loewy's classroom. He immediately bought into American consumer style, where any item needed was readily available. "My French friends can't believe, even in a small village, one can send Junior for an ounce of mercurochrome, two records by Spike Jones, an automobile jack and a pint of chop suey," he wrote. Still, he felt stifled despite his success, and considered returning to France to try his hand at farming (admittedly, it's hard to imagine Loewy straddling a tractor or herding sheep) or clothing design. Still, the abject wonder Loewy felt at the array of consumer products did not prevent him from criticizing their design. He felt most American products were overdecorated, full of "gingerbread, spinach and schmaltz." In addition, this "spinach and schmaltz," which became Loewy's favorite descriptors to characterize bad decorative design, made goods more expensive for the consumer. Much of the decoration for retail products had to be added on in a separate process, whether it was painted, etched, embossed, or stamped. In his opinion, manufacturers underestimated the taste of their customers.

The customer base Loewy was speaking for was rapidly growing in the years after his arrival in 1919. The American economy was undergoing a colossal industrial expansion. The country's manufacturers were switching from craftsmen working on a single product to assembly-line manufacturing that allowed firms to produce many more items yet charge lower prices. Consumer wages were rising, allowing families' discretionary income to buy these less expensive products. At the same time, electrification was expanding beyond the East Coast, and families were migrating from rural agricultural jobs to manufacturing jobs in urban centers. Roadways, many of which were little more than dirt roads that turned to muck in the winter, were developing, too. The first paved highway, the Lincoln Highway (known today as Route 30), stretched in a meandering route from New York to San Francisco. Trucks had replaced horse-drawn wagons for city deliveries, and larger trucks were starting to be used for interstate commerce. There were 2.5 million miles of road stretching across the United States, but less than 10 percent were paved with any type of improvable surface.

The changing atmosphere of America after the Great War showed Loewy that a fashion illustrator did not necessarily have to remain one. He was making $30,000 to $40,000 a year (about $370,000 today), according to his writings of the era, but drawing ads did not seem to him to be vital work. "I felt I belonged here [in the United States]. I liked everything about it. How could I leave all this forever? It seemed impossible."

He prepared a business card bearing the message "Between two products equal in price, function and quality, the better looking will outsell the other." Loewy sent the card and an introductory letter out to all the contacts from his illustration clientele. He decided to hit the road to pitch his services to manufacturers and transportation executives.

The tug of his engineering education protected against complacency. As part of his illustration work, Loewy branched out into doing assignments and ad illustrations for advertising agencies. One of the firms with which Loewy established a solid working relationship was the New York–based Foote, Cone and Belding, one of the first "multiple principal" ad agencies (leadership was a partnership between several principals) and one of the first to offer "public relations" as a service beyond creating advertisements and slogans. One of the agency's clients was Sigmund Gestetner, a British manufacturer whose products included early versions of office duplicating machines.

Loewy described his first meeting with Gestetner as being much more exotic than meeting in the waiting room of an ad agency. He recalls in *Never Leave Well Enough Alone* that he met Gestetner in London when the businessman overheard Loewy analyzing the functional superiority of English taxicabs over American ones. In many of Loewy's client-pitch stories, the designer makes a new acquaintance by offering his opinion of one product over another. The truth of their meeting is probably a mixture of fact and fantasy, but one part remains true: meeting Sigmund Gestetner in 1929 changed the path of Loewy's career.

Gestetner asked Loewy a simple question: "Can you improve this machine?" Loewy did not hesitate to name a price, a boldness that would seem to be foolhardy, as he had never designed a product before, aside from his *Ayrel* model airplane. Never at a loss for the bold gesture, Loewy informed his customer that he could reimagine the duplicating machine in three days, offering to charge only $500 instead of $2,000 if Gestetner did not like the design. The office equipment and duplicators marketed by Gestetner had many moving parts, almost all of which had the potential to snag the clothing of the employees operating the machines. The machines were efficient, but their look was dictated by functionality, not aesthetics. After Gestetner agreed to Loewy's estimate, the fledgling designer writes that he celebrated the beginning of a new career by opening a bottle of Moët champagne. He also immediately ordered 100 pounds of modeling clay, tools, and a floodlight.

The duplicating machine "looked like a very shy, unhappy machine. It smelled—smelled of oil, ink and leather. It wasn't nice." Characterizing his work as amputation and plastic surgery, Loewy applied the modeling clay directly onto the machine. The amputation refers to Loewy's decision to eliminate the long legs of the machine's platform and to create a storage cabinet with much shorter legs beneath. The plastic surgery was literally true. Loewy simply created a removable shell of Bakelite that encased all the machinery inside the shell. The duplicator that Gestetner brought to the studio resembled nothing so much as one of the original designs from the dawn of the Industrial Revolution. Its inner workings were all exposed, with various pulleys, tensioners, clamps, and bars providing ample surfaces for dust, grime, and grease. The storage cabinet was lashed between tubular legs that splayed outward from the line of the cabinet, a design almost guaranteed to trip at least one office worker per day.

Loewy, who loved to celebrate his Gestetner redesign in his presentations, described the machine as being "covered with a mysterious bluish down that looked like the mold on tired Gorgonzola." Loewy often said that the original design lasted forty years without a change, which meant that the design was perfect or that Gestetner was too frugal to pay for another makeover. In truth, the duplicator was not a popular consumer product and did not require periodic design changes. "I often kidded Sigmund about the fact that an exceptionally successful design can make a fortune for the client and put the designer nearly out of business, waiting forty years for the next assignment." Loewy redesigned or originated designs for other equipment for the company throughout his entire career.

Loewy's redesign of the duplicator was a masterpiece of repackaging. It featured a smooth, architectural shape, perched on an elegant top with rounded corners. The splayed legs were shortened and incorporated into the line of the cabinet. The duplicator's machinery was cloaked by a casing that revealed only those controls needed to operate it. The objectionable smells and inks were mostly confined inside the casing, and dust was barred from most of the working mechanisms. The designer

Raymond Loewy's first assignment as an industrial designer. Sigmund Gestetner, a manufacturer of office machines, asked Loewy to make over a duplicating machine. His elegant solution was to encase most of the machine in a handsome streamlined shell. Loewy hired a clay modeler to help him conceive the final look. (Raymond Loewy™/® by CMG Worldwide, Inc./www.RaymondLoewy.com)

had hired help, but the conception and editing of the final design were obviously his. Loewy presented the finished design to Gestetner. Loewy not only won his wager when the industrialist accepted the design and paid the $2,000 fee (roughly $28,000 in 2015), but also accepted a retainer from the company to design other products. Aside from his commission, however, Loewy's big breakthrough brought him little fame and less publicity. Gestetner would not put the redesign into production until 1933. Loewy still had to market himself relentlessly to build his reputation for years after his first major design. As it was, Loewy pleased his client and increased sales for the company, which became a central mission for his own enterprise. If he accomplished nothing else, that legacy was praise enough for the designer.

The facelift given to the Gestetner duplicator revealed the gifts and weaknesses in Loewy that competitors and critics would alternately praise and criticize. The European design ideal, taking its inspiration from the Arts and Crafts movement of William Morris and later the Bauhaus designers, preached finding the purest design, reducing a product to the perfect expression of utility. Loewy, the Parisian, preferred to trust his own taste and apply it to uniquely American products.

The truth is that Loewy never trained in aesthetics or art school theory. Conscripted essentially straight out of school, Loewy did not have the time to absorb the intricacies of form and function. He found inspiration in the goods (at least in New York City) seemingly available to anyone willing to work hard enough to buy them. His judgment was formed after he came to America and wandered the streets to market himself as an artist. His fondness for beautiful products was certainly informed by his European years, but his ability to see what would sell to customers was uniquely American. Design historian Philippe Tretiak captured Loewy's philosophy succinctly: "He adopted the marketing philosophy of his country." And Loewy's most remembered maxim says it all: "The goal of design is to sell. The loveliest curve I know is the sales curve."

Raymond Loewy was not designing in a vacuum. The output of salable products, created in factories and often designed by engineers or mechanics who prized workability over style, was so large that companies sought experts who could make their item stand out in the marketplace. It did not matter what type of experience this new breed of designer had. It mattered only that they could create a need to buy. Designers also emerged from backgrounds in theater, academia, graphic design, and fine art. Before establishing their own businesses, many designers also were associated with advertising agencies, either as freelance specialists or as employees.

Although the Great War ravaged an entire generation of young European men, many early industrial designers were European immigrants. Some came to America before the war to avoid being sucked into the draft, and others, like Loewy, arrived after completing their military service. Kem Weber came from Germany, Paul Frankl was born in Prague (part of Austria-Hungary), and Joseph Urban came from Austria, followed by John Vassos from Greece and Paul Lazlo from Hungary. French sociologist André Siegfried wrote in *America Comes of Age*, "The American people are now creating on a vast scale an entirely original social structure which bears only a superficial resemblance to the European. It may even be a new age." These émigrés were artist-illustrators who crossed over to industrial product design. From 1918 to roughly 1925, designers from disparate professions were trickling away from their first professions into a new paradigm. The wedge with which some designers entered the profession was package design. The primary vehicle for selling products, particularly national brands, was standardized, familiar packages. The packages in turn achieved familiarity through designed national magazine ads.

The early advertising agencies branched out into other services, and product design soon was established as a subset of agency services. National agencies were quick in gaining unique insight into the mind of the American consumer. One pioneering agency created a manual, called the *J. Walter Thompson Book,* describing the American buyer this way: "He lives in an atmosphere of action. He is always looking for new ideas; he appreciates improvements and inventions; he understands the value of time, and of taking shortcuts to get what he wants."

As agencies were consolidating, the standardization of machine parts also increased efficiency within the new manufacturing centers. Rising wages and more efficient workflow meant that workers had more to spend and had a bit more time to do so. Industrial production figures doubled between 1919 and 1929, while the automotive industry exploded, climbing from one car for every 184 consumers in 1910 to one car for every 5 consumers in 1930. The first concrete examples of how mass production and mass advertising affected the country's consumer base came through the auto industry. It was Henry Ford who had the better idea, at least where efficient manufacturing was concerned. The Ford assembly line democratized the ability to purchase a family car. This innovation would lead to the theory of consumption known as Fordism, which, in part, preaches that workers must be encouraged to consume the very products they produced.

Americans also had more free time to use the goods they aspired to. The work week decreased almost 10 hours (to 50.6 hours) from 1900 to 1926. According to Gary S. Cross in *An All Consuming Century: Why Commercialism Won in Modern America*, "Consumer goods were the building blocks for the construction of different identities and new communities when the old ones were in decline." As people decided what new identity to strive for, they sought out suitable items fit for that lifestyle. "The car was the bellwether commodity of the new century."

Differentiating products in the marketplace became one of the most pressing business problems of the day. While advertising effectively communicated the advantages of owning a particular car and succeeded in getting customers to the dealership, mere ads could not seal the final purchase. Carmakers had to distinguish models through something other than marketing. Detroit and the independent carmakers soon marketed vehicles based mainly on how a car looked. Customers had to react to a product in ways that reflected how they perceived themselves. The saturated market for consumer goods made it imperative for companies to bring in designers or "stylists" to make their products stand out. The term "industrial designer" was first used in 1919, but initially companies turned to stylists on staff with advertising agencies. Some companies hired designers as in-house staff, such as Ray Patten, who was hired in 1928 by General Electric to design appliances. Westinghouse hired Donald Dohner shortly after that to design its own appliance line.

At the same time, artists and designers in other professions saw the opportunity to branch out from their more narrowly defined specialties. One of the first to make the leap was Egmont Arens, who joined the art department of the Calkins and Holden ad agency. He specialized in package design, which became an early area of agency specialization because single-product package work required little or no support staff.

Another Calkins alumnus, Walter Dorwin Teague, came into the profession with fifteen years of experience as an advertising illustrator. Teague was one of the first industrial designers to form his own multiple-client company. He and Loewy were also among the first designers to delegate most of the design work to employees while concentrating on bringing in business. Teague's first major client, signed in 1927, was Eastman Kodak, a client he would keep for three decades.

Teague came from a deeply conservative upbringing as the son of a circuit-riding Methodist minister in Pendleton, Indiana. Born in 1883, he left Indiana for New York City to study art at the Art Students League, where he supported his education by painting signs and illustrating mail-order catalogs. By 1908, he had joined the art department at Calkins and Holden. Teague loved classical forms, particularly motifs from French culture such as fleur-de-lis. His gift for design came not in original motifs, but rather in recombining forms, often historical or classical images, in new ways. Teague was never a prolific originator of design ideas, and he turned the design work over to employees fairly early in his career. One of the most revealing photos of what an industrial design office should look like came from Teague. A photo in Arthur Pulos's *The American Design Adventure* shows Teague in the foreground at a large drawing table. In the background are two rows of smaller drawing tables receding into the distance, each desk manned by nearly identically dressed designers. The image reinforces Teague's idea of a well-regimented design office and functions as a satiric comment on the postwar idea of the "organization man."

While Loewy's first commissions were completed with a minimum of personal marketing and aggrandizement, another seminal industrial designer destined to become a Loewy contemporary cultivated the image of the genius artist. Norman Bel Geddes was born in 1893 in Adrian, Michigan. He was brought up under Christian Science, which preached the power of mind over matter and thus gave him an unshakeable faith in his own talent. He started his studies at the Cleveland Institute of Art, but left to live for a summer on a Blackfoot Indian reservation. He returned to study at the Chicago Art Institute but dropped out owing to lack of funds. Bel Geddes went to work in Detroit, starting at Peninsular Engraving Company designing ad posters and theater program covers. He later freelanced at Detroit advertising agencies, working primarily on automotive accounts rendering "beauty shots" of car models using an impressionistic technique. As he worked on more auto accounts, he began to specialize in model building, an interest that led him into theatrical stage design when he convinced Aline Barnsdall to let him design the stage of Los Angeles Little Theater. Bel Geddes worked with Frank Lloyd Wright, who was designing the theater building. Later accounts of the collaboration say that the clash of two monstrous egos doomed the effort to failure before it started.

He even designed his own persona, adding the "Bel" to his last name to give him an aristocratic air. Norman Melancton Geddes co-opted the middle name his of first wife (Helen Belle Schneider) to create a more artistic identity. His designs for stage productions were literally monumental. He designed an onstage ship launch by moving an entire mockup of an ocean liner on ball bearings. For the 1923 play *The Miracle*, he turned the entire interior of the theater and the stage into a medieval cathedral. Before hanging his shingle as a designer, Bel Geddes worked for Universal Studios, writing and directing *Nathan Hale* in 1917 and designing *Feet of Clay* for Cecil B. DeMille and *Sorrows of Satan* for D. W. Griffith. He also developed high-magnification photography for the nature documentary *Amphibia and Reptilia*.

Like Teague and Loewy, Bel Geddes expanded his portfolio by taking on commissions for store window displays. As Bel Geddes put it, "The window is a stage, the merchandise as the players and the public as the audience." He originated window displays for the store for two years and eventually hired a woman, Frances Resor Waite, to oversee the account. He married Frances, after divorcing Helen Schneider, in 1933, and she introduced Bel Geddes to her uncle, Stanley Resor, president of the J. Walter Thompson advertising agencies. Resor had climbed to the top of the agency after Thompson abdicated most of his duties and the agency had slid badly in the absence of leadership. Resor's talents were creating memorable brand names and positioning products as upper-class accessories that were available to the middle class. His son-in-law Bel Geddes tended to fail upward, usually by alienating clients who asked him to design a product and then watched slack-jawed as the designer delivered a grandiose plan for not only the product but also the manufacturing line and perhaps even the trucks to be used as delivery vehicles.

Bel Geddes gave Loewy a run for his money as a self-promoter, but he lacked Loewy's personal reticence. In other words, he didn't know when to leave well enough alone. His early designs included a model bedroom suite for Simmons, a sizzling design in metal and black lacquer that sold poorly, and a sleek grocery scale for Toledo Scale that never quite made it into production. A Toledo Scale executive characterized Bel Geddes as "The man who when asked to design a product, went on to design its factory."

Perhaps the most influential designer to emerge in the early days of the profession was a protégé, or more accurately an understudy, to Norman Bel Geddes. His name was Henry Dreyfuss. He was less concerned with how products looked than he was by how people used them. He would go on to fine-tune this interest into the science of ergonomics, of adapting products to be the most efficient in a working environment. Almost the opposite of Loewy in temperament, the low-key Dreyfuss recognized the value of flying under the radar. He wore a uniform of sorts for himself, a plain brown suit, so as not to outshine the clients who came to him seeking a design solution. Furniture designer George Nelson said, "He seemed immersed in the establishment, a terrible square always taking the corporate side in any argument."

Dreyfuss was rarely interested in styling a product to increase sales, although he certainly was successful at it. Instead, he concentrated on finding out how people related to the product. He told colleagues, "Make machines fit people—don't squeeze people into machines." The official credo of the Dreyfuss office (which was perhaps the only major design office to have an official credo) was "What we are working on is going to be ridden in, sat upon, looked at, talked into, activated, operated or in some way used by people individually or en masse."

Dreyfuss's grandparents had emigrated to America from Mannheim, Germany. Henry's grandfather, Moritz, founded a small company creating theatrical costumes and scenery. His father, who died when Dreyfuss was 11, worked as a tailor in the family's theater supply firm. Young Henry was interested in the arts nearly by default. After high school, he toured Europe and returned to serve an internship with theater designer Bel Geddes. He considered Bel Geddes "the only authentic genius this profession has produced." Macy's, always looking for up-and-coming window dressers, offered Dreyfuss an office job focused on identifying store merchandise for possible redesign. He turned the position down, saying, "An honest job of design should flow from the inside out, not from the outside in."

Dreyfuss, in contrast to Loewy, was loath to overshadow the client. In addition to wearing his famous brown suit, he could draw upside-down—a skill that helped explain concepts to clients seated on the opposite side of a table. He felt that the guiding principle of design was satisfaction in product use. "If people are made safer, more comfortable, more eager to purchase, more efficient—or just plain happier—the designer has succeeded." He did not feel comfortable putting new clothes on an old design. For client after client, he preferred working with company engineers and sales staff to design a product from scratch. He was the only American designer of the profession's pioneers to not mine the art and design styles of the day. His telephone designs were simple, unadorned, and functional. More than any other designer, Dreyfuss approached the modern ideal of form follows function.

The use of ergonomics grew out of Dreyfuss's encyclopedic research on the human body. He kept line drawings of the human body in male and female form in his New York and California offices. Calling them Joe and Josephine, the charts noted average body measurements, which he supplemented with myriad computations, angles, and figures. Soon he had amassed average measurements for hundreds of body positions. He also paid attention to what users said about the products he was designing. One of the firm's first commissions was the design of a Royal typewriter. Secretaries using the product had complained of eyestrain and headaches. Dreyfuss's solution was to eliminate the product's black lacquer finish, which was reflecting the overhead lighting, and use instead a dull, pebbled finish. It may not have gleamed like a new car, but it was certainly easier on users' eyes.

"Ours is the ever-changing battleground of the department store rather than the Elysian Fields of the museum," he wrote. Dreyfuss also believed in testing and retesting. He truly pioneered market research specialized for industrial design. Dreyfuss washed clothes, cooked, drove tractors, operated diesel locomotives, spread manure, drove a tank, and vacuumed rugs. Research was his religion.

No fan of the streamlined look, Dreyfuss felt applying streamlining to an object that would never be in motion was stunt designing. He preferred to call his designs "cleanlining," a rebranding that appealed to most of his competitors, all of whom wanted to avoid being grouped into a singular style. A sharp example is in the design for Mason jars. Dreyfuss simplified the form by making the jars square with tapered sides. The design allowed stores to get more on a shelf and prevented the jars from rolling off flat surfaces.

Dreyfuss's devotion to ergonomics ultimately meant that his designs were longer-lived. Indeed, some have never gone out of style. The Honeywell circular thermostat, designed in 1941, is still a popular product of its kind and is certainly the easiest and most intuitive to operate. His work for Bell Telephone had similar lasting effects. The original cradle telephone was nearly perfect in its design and execution. The handset is easy to hold, the ear- and mouthpiece work for both men and women, and the proportion between the handle and the ear and mouthpieces is perfect for using a shoulder to hold the handset against the ear.

LOEWY, DREYFUSS, TEAGUE, and Bel Geddes all came to the attention of American consumers and manufacturers within a few years' time. At times, these designers were polite rivals, making snide remarks about taste and work habits. Bel Geddes once said of Loewy, "He was more interested in living than designing. He would take summer drives in the biggest white convertible with the most beautiful blondes—he just looked like a designer."

The new vibrant products of the era became known as "style goods," referencing the options available in two stylish businesses: women's fashion and furniture. In some cases, manufacturers used package design to create a new image. In other cases, they redesigned how the product looked. In both cases, companies began to search out men and women who could bring style to objects that previously failed to attract customers. Advertising agencies soon created design departments as part of their services. By 1931, Young and Rubicam Agency had renamed its art director as vice president in charge of design. Earnest Elmo Calkins essentially saw his advertising agency as a preservation society for style and beauty, saying that workers could not add beauty to their products because their machine-age jobs prevented them from adding beauty through their own craftsmanship. "If we are to have beauty in the machine age, it must be imposed from the top by the fiat of the man that owns the machines."

Raymond Loewy's early commissions in his first career as an illustrator came through advertising agencies, either as a freelancer or as the "house" artist for a manufacturer or retailer. Loewy did extensive work for Lord and Thomas Agency (later Foote, Cone and Belding) and Lennen and Mitchell Agency. Lord and Thomas was run by legendary ad man Albert Lasker, one of the few Jewish men to run a major advertising agency in the early days of the business, and Loewy and Lasker became acquainted socially. Loewy in turn hired Mary Reinhardt, a socially connected woman (who would marry Lasker in 1940) as his public relations representative. She helped introduce Loewy to clients such as American Tobacco (a Lasker client). Once he established his design business, Loewy was permitted to mention to potential clients that his new design firm's staff had access to Lord and Thomas's research department. Although Loewy never worked at an ad agency as an employee, his connection to Lord and Thomas had a major impact on his early success.

Loewy's initial commissions through ad agencies gave him a model on which to base his own business. He also was witness to how other freelance professionals, such as public relations impresario Benjamin Sonnenberg Sr., created an image. Sonnenberg, who wore a pseudo-Edwardian getup of Chesterfield coat, homburg, spats, and a cane almost every day of his working life, used his public persona as a calling card. New York businessmen recognized him from blocks away. Loewy saw the value in dressing as a European sophisticate, and always appeared in bold but tasteful suits. The ad agencies of the 1920s touted research, but in general such research was limited to determining geographic buying power, noting brand preferences at targeted stores, and studying media coverage patterns. Agencies made attempts at plumbing the mind of the consumer, but most research tracking consisted of testing ads or samples at haphazardly picked groups or sending questionnaires to a group of friends. The industrial designers, save for the painstakingly scientific approach of Henry Dreyfuss, used similar research methods, and as the ad industry refined its methodology, the designers updated or improved their own.

One area in which Loewy did little or no research was how to market himself. Intuitively, and later with the help of public relations professionals, Loewy understood that he should present an image that logically underlined his expertise as a product designer, and Loewy admitted he made living well part of his business persona. "I'm sure my lifestyle was an easy target for other designers. It probably made some of them resentful and critical, because luxurious living didn't seem to interfere with the firm's increased reputation and large output. But a good life has been as important to me as my work; in fact, the two of them are bound up in each other for me."

The beginning designs of the new professional industrial designers were just the start of a new era in consumer marketing. Nearly every industrial designer who found clients in the initial days at the beginning of the 1920s and 1930s went on to create design firms of their own. Many kept things small, accepting just enough commissions to make incremental expansions. Only a few, however, had the vision and the gambling spirit to create a coast-to-coast operation. That was left up to Teague, Dreyfuss, and Loewy.

3

The Artist (and Others) Shape the Things to Come

Norman Bel Geddes lived to issue manifestos, and he truly believed that he could change the world through his ideas. He nearly did it by publishing a slim book filled with a short essay on design and many drawings of streamlined objects, most of them focused on modes of transportation. Automobiles, trains, ocean liners, and airplanes are on display throughout *Horizons*, which debuted in 1932. It was a manifesto for a new style of product design—streamlining.

The ideal shape for streamlined design was the teardrop, an idea reflected repeatedly in cars, airplanes, and most memorably in a Raymond Loewy–designed pencil sharpener that never made it to production. Loewy created the one-off in 1933 and exhibited it in 1934 at the Industrial Arts Exposition in Rockefeller Center. Loewy went so far as to declare a patent for the design, but he could not sell it to a manufacturer. Ironically, it may be the most famous design never to have been put onto production. In June 2011, the US Postal Service created twelve "Forever Stamps" featuring the work of industrial designers. Loewy's pencil sharpener was included—the only commemorated design never to reach the mass market. Industrial designer Budd Steinhilber recalled the famous design, which he used as a young Loewy intern in 1942. "One of my daily morning task assignments was to make certain that all of the pencils in Raymond's office were sharpened to a sharp point. This shiny model was proudly displayed on his side table," Steinhilber recalled. "But I discovered it didn't sharpen pencils worth a damn. I used a trusty old Boston pencil sharpener instead. My second discovery was that on the following mornings none of the pencils had been used (which I surmised, spoke to the frequency with Mr. Loewy did any sketching personally)."

Harold Van Doren, the designer responsible for Toledo Scale models and Maytag appliance designs, zeroed in on Bel Geddes's popular appeal while slyly commenting on the designer's tendency for the superficial.

The "teardrop" pencil sharpener created by Raymond Loewy may be the most celebrated design never to go into production. Loewy took out a patent application for the design, and in 2011, the US Postal Service used it on one of their "Forever" stamps featuring the work of industrial designers. "It didn't sharpen pencils worth a damn," recalled former Loewy designer Budd Steinhilber. (Raymond Loewy™/® by CMG Worldwide, Inc. / www.RaymondLoewy.com)

"Everything that moved through air or water was manipulated into the shape of a catfish." Later, when one of his streamlined car designs was turned down for manufacture, Bel Geddes wrote somewhat torturously, "The car was never built, owing to psychological factors in the human makeup having to do with timidity." There was absolutely no timidity in Bel Geddes's manifesto. He proposed that all transportation products eliminate all projections and flat surfaces. The snake-headed *Locomotive No. 1* is totally encased by its outer skin, an impractical but wind-resistant design, yet the ultramodern *Lounge Car No. 4* is so contemporary that it resembles today's "bullet train" locomotives, particularly the French TVA engine and Amtrak's *Acela*.

Horizons contains Bel Geddes's trademark "big picture" designs. Never content to design a single product, he loved creating systems. His "Air Terminal" is laid out in an immense square with four runways that allowed multiple planes to land simultaneously. It featured a massive underground repair facility, and in an anticipatory flash of insight, a moving sidewalk for weary commuters. Not content with this design, he put forth a "Rotary Airport," an S-shaped terminal floating in New York Harbor that could pivot in any direction to accept planes. The seaplane, which was just becoming popular through Pan American Airways' Clipper aircraft, also found a champion in the visionary designer. "Airliner No. 4" is one of the most ambitious follies ever committed to paper. A gigantic flying aircraft powered by ten propeller engines, the design featured passenger space, complete with windows, in the wings and mammoth pontoon floats. With utter certainty, he wrote, "It will fly much more smoothly than any plane that has yet been built, if for no other reason than its enormous size." The aircraft was planned to carry 606 people—451 passengers and a crew of 155. It featured a 528-foot wingspan, and its pontoons were 235 feet long and 60 feet high. Had it been built, it would have made Howard Hughes's *Spruce Goose* look like a Piper Cub. Design historian Christopher Innes calls the plane's design "certainly practical" and credits it with inspiring the *Spruce Goose.*

The popularity of *Horizons* made Bel Geddes the first superstar designer, despite his spare track record of usable designs. His aura of authority over all matters of design soon extended to the rest of the newly minted artist-designers of the age: Loewy, Teague, Dreyfuss, and to a lesser extent, Van Doren. Innes makes the case in *Designing Modern America* that Bel Geddes was the father of modern American design (along with architect Joseph Urban). And while it is true that Bel Geddes conceived and publicized streamlining before his contemporaries, the inescapable fact is that many of the products he designed were never produced.

Although Bel Geddes filled *Horizons* with all manner of self-aggrandizing pronouncements, he did get his final thoughts right. "All the industrial design we have had in the United States, as yet, is comparable in effect to a pebble dropped into a pond. The circles that have agitated the surface will continue to widen and spread with an ever-expanding sphere of influence."

Many of the designers who competed for clients with the bombastic Bel Geddes emerged after World War I. Almost all of them were fond of declaring that the profession of industrial design emerged full grown from their collective talents. In reality, Americans had been designing products since the first settlements in the seventeenth century. Many of the tools and implements used in these early settlements were imported from England, France, or Spain and adapted for the settlers' own needs or circumstances.

A truly unique American design to emerge—from the coach-building trade—was the Conestoga wagon. Built to withstand the epic journey across the American West, these prairie schooners, many of them built by the Studebaker Company, were perfectly realized products designed to deliver families to a new future. It was the original recreational vehicle, save for oxen instead of an optional hot tub.

A series of American inventors—such as Oliver Evans (1755–1899), who developed an early steam engine, and Robert Fulton (1765–1815), who adapted a steam engine for use on a boat—brought a manufacturing economy to the country. Inventor Eli Whitney's (1765–1825) cotton gin allowed cotton plants to be cleaned quickly and economically. Within a few years, the textile industry had begun to prosper, making cotton farming economical in the South and starting the manufacture of cloth in factories in the North. Whitney also proposed a system of standardized interchangeable parts for producing muskets and other firearms.

The Conestoga wagon that settled the West also was supplanted by eastern canal systems and a growing railroad system, which eventually supplied a nationwide distribution network for a growing list of products. The railroad soon became a repository for newly designed machines. The Pennsylvania Railroad Company was founded in 1823, but its first locomotives were English. The Pennsy soon found that bigger, more powerful engines were needed for the country's wide-open spaces. The first great locomotive designer, M. W. Baldwin, started his career as a jeweler and was later assigned to build a miniature engine for a Philadelphia museum. Much as coach making defined the previous century and aircraft would define the following century, the locomotive became visual shorthand for nineteenth-century technology. These design breakthroughs were complemented by the creation of standardized iron parts, a system of building that would eventually allow larger and taller buildings to be constructed.

Starting with the 1851 London Exhibition, international expositions became the vehicle for showcasing the latest works from the host nation's factories. This tradition persisted well into the twentieth century, culminating in influence with the 1939 World's Fair in New York City and sputtering out after the 1964 World's Fair, also in New York.

For America, the Centennial Exhibition of 1876 in Philadelphia was the most successful of the early American exhibitions. More than 10 million visitors trooped through the displays, and most of them left realizing that many of the products before them were affordable versions of products previously associated with the privileged class.

At this point, the purchasing public was not concerned with aesthetics or form following function. They were more interested in making their lives easier. The abundance of objects designed to improve citizens' lifestyle included apple peelers, meat grinders, coffee mills, and flour sifters. Two of the most transformative home products of the nineteenth century were the cooking stove and the sewing machine. The stove became the focal point of the kitchen and eased the unreliability of other cooking methods. The sewing machine, saving hours of needlework, turned women into a new class of worker. It industrialized the clothing industry and created jobs for poor immigrant women. As the price of clothing dropped, the durability of clothes became less important to consumers, and style and fashion became more important.

The explosion in consumerism was simultaneously fueled by a population increase from 25 million to 75 million between 1850 and 1900. Many of these new product buyers were immigrants, who quickly (often through employment at the expanding factories) joined the lower middle class, often within a generation. During this same period, American jobs would shift from primarily agricultural to industrial careers. To spread the word about new products and to tout upward mobility, the relatively new media industry expanded. Newspapers multiplied as morning and afternoon dailies competed for advertising dollars. Magazines rapidly increased circulation as well, from 1,200 publications in 1870 to 2,400 in 1880. The rise of advertising agencies fed the need to offer guidance to manufacturers seeking to market wares directly to the public.

Around the 1880s, the nation's expanding middle class began to separate into divisions, led by businessmen who had found fortunes in retail, transportation, manufacturing, and other endeavors. These upward strivers did not look toward homegrown enterprises to purchase objects to outfit their newly constructed palatial homes. Instead, they traveled throughout Europe, took the Grand Tour, bought aristocratic titles, sent their children to elite boarding schools, and, if they were lucky, married into royalty (preferably monied), however obscure. This time, according to design historian Arthur J. Pulos, found the "wealthy eastern establishment and its hired curators of eclectic taste impos[ing] the fashionable French Renaissance, Italian Romanesque and Greco-Roman classical styles on the United States."

These curators of eclectic taste were architects such as H. H. Richardson, Richard Morris Hunt, and Louis Sullivan, all of whom were involved in the displays of the United States' next great industrial showcase, the Columbian Exposition of 1893. The relatively new phenomenon of electrification revealed another array of appliances promising to make American lives easier. Electrified teakettles, griddles, coffeepots, and heaters were put on display, conveniently enough, in the Electricity Building. The building housed an entirely electrified kitchen and Thomas Edison's Kinetoscope. The exhibition also provided a first look at a product style that was a sharp turn away from the ornate neo-European styles that had dominated the look of architecture and product design.

The Arts and Crafts movement seeped into the cultural consciousness of the middle class through the architecture of Frank Lloyd Wright and other practitioners of the Prairie School. Often these Prairie homes came with furnishings designed specifically for the home, and often the supplied Wright furniture was much more satisfying to look at than to sit on. The movement reached popularity with the plain products—particularly simple oak chairs—of the Roycroft shops, an enterprise that began as a publishing press and morphed into bookbinding, leatherwork, and furniture. The tipping point came in the form of Mission Style furniture by Gustav Stickley and the furnishings of Louis Comfort Tiffany.

The turn of the century seemed to promise an ever-expanding horizon of technology. The supply of oil products and electricity seemed infinite, and American manufacturers and citizens responded with "a frenzy of inventions based upon gasoline engines, electric motors and heating coils." Yet another exhibition, the 1901 Pan-American Exposition in Buffalo, New York, effectively built a shrine to electrification. Everything from the buildings to the foliage and fountains were illuminated with electricity acquired from the largest power plant in the world, in nearby Niagara Falls. From 1880 to 1895, the country saw the introduction of the electric- and gasoline-powered car, the electric trolley, and the bicycle. This was the world in which Dreyfuss, Bel Geddes, Teague, Van Doren, and another emerging designer, Donald Deskey, formed their design aesthetic. Loewy would see much the same developments in his formative years in Paris.

Gradually, products began to evolve into distinctive forms. Early in the new century, however, function took precedence over form. In fact, form was lower in priority than almost any other aspect of the product. In describing the look of the new consumer society, Pulos said, "Automobiles were horseless carriages, parlor or cooking stoves were baroque idols or altars, cash registers were Renaissance jewel safes, lamps were Art Nouveau gardens and furniture was still lost in Victorian reverie." Manufacturers' aesthetic decisions were based on what moved the merchandise and what could easily be adapted to an always quickening production line. Although Europe had established design collectives such as the Deutscher Werkbund—which would produce or influence such architects and designers as Walter Gropius, Ludwig Mies van der Rohe, and Le Corbusier—Americans were slowly catching up.

Design in the United States came to the fore through the invention and adaptation of mass production. Nowhere was mass production more easily understood than in the auto industry. The process of creating a single automobile from parts manufactured off-site was not invented or even adapted early by Henry Ford. Instead, Ransom E. Olds, the automotive entrepreneur whose REO brand inspired early American truck design (and the 1980s band REO Speedwagon) and whose later company, Oldsmobile, was absorbed into the automotive conglomerate General Motors, started mass production in his factory in 1899. Henry

Ford's insight was to refine mass production to the point of religious vocation. By 1913 and 1914, Ford had streamlined the process of building a Model T at the Ford Motor plant in Highland Park, Michigan, by standardizing parts and giving workers specialized roles. The time of assembly for a Model T was reduced from 12.5 hours to just 1.5 hours.

Mass production also gave manufacturers the opportunity to establish new forms. The introduction of large presses to shape sheet metal allowed companies to shape their automobiles using sloping fender lines and, later, unified bodies. The previous method of body shaping involved hand-hammering metal sheets over wooden forms, a process inherited from carriage making. The increased sales from mass production lines showed businesses that corporate logos and packaging were instrumental in maintaining customer loyalty. Henry Ford made sure the script form of his name was put on every car leaving the line, and he renewed the trademark faithfully. Coca-Cola trademarked its script logo, which was created by Frank Mason. The Coca-Cola bottle also was redesigned to make it instantly recognizable on a store shelf. The "hobble skirt" bottle, developed in 1916, was designed by Alexander Samuelsson, a Swedish glass engineer who won a company design competition. Raymond Loewy, particularly after he was a successful designer, would often start invited lectures on industrial design by pointing out that the Coke bottle was an example of perfect design, high praise from the famous designer. His praise also gave the impression that he was responsible for the design, a misunderstanding that persists decades beyond Loewy's death.

American design and the manufacturers using design services escaped the devastation of World War I, which proved to be a beacon for European designers seeking to put an imprint on US products. Most of Europe underwent economic hardship and political upheaval, but the design émigrés were not necessarily fleeing their homeland. Instead, they were part of a quest for opportunity. The newly arrived artisans would represent a major portion of the first generation of professional industrial designers, carrying with them the influences of the movement and the eclectic product styles of the European marketplace.

Designers such as Walter Dorwin Teague and Egmont Arens began their careers at advertising agencies, as did Donald Deskey. Deskey's career evolution is typical of the pioneer designers. A native of Blue Earth, Minnesota, he began at a Chicago agency and moved to New York in 1921.

But he was fired within a few months because the agency considered his work too cutting edge. Rather than retreat, Deskey opened his own art agency and was soon earning $12,000 a year (roughly $140,000 today). The young designer's uneasy career plans led him to close the agency to study the new art being shown in Paris. He returned to the United States in 1922 to tiny Huntingdon, Pennsylvania, where he accompanied his wife, Mary Douthett, a music instructor, to teach at Juniata College.

"The mystery is that this particular man should have ever have designed anything at all, for until he was nearly 30, Deskey's connection with art was either that of an amateur or of an advertising man, and the future he looked forward to was that of a sound and energetic American man of business. If he had not been by nature something of a tramp, he might have been chosen to do advertising for [Radio City Music Hall] instead of designing." These words, written by Gilbert Seldes in a 1933 *New Yorker* profile, portrays Deskey as the archetypical self-made designer. In 1925, the Decorative Arts Exhibition drew him back to Paris. When he returned to the United States a year later, he found that his designs for screens and other furniture were sought out by architects and theatrical designers. He soon opened his own firm in New York City, where his eye for streamlined furniture design and eye-catching materials attracted style-conscious and extremely rich clients, including John D. Rockefeller (more accurately, Rockefeller's wife, Abby).

This career scenario was repeated over and over, with minor variations, in the post–World War I economy by almost every designer carrying a mechanical pencil and a drawing board. Whether an American such as Bel Geddes and Dreyfuss, or a recent arrival such as Loewy, Vassos, or Weber, it seemed as though anyone with an iota of art training and a bold eye or opinion could establish a design business and bring in eager clients. Other designers emerged from occupations on art's periphery. Harold Van Doren came into design from a career as a museum director. None had advanced degrees in engineering or architecture, and few came from prestigious academic programs steeped in art and design, because few academic design programs existed at that point. According to Stephen Bayley and Terence Conran, "Designers and manufacturers were always aware of the demands of the marketplace and were quite unashamed to respond to them."

At the onset of this convergence of talent, a new art movement from Europe was conceived at the Paris Decorative Arts Exhibition. The exhibition popularized the style that later would come to be called Art Deco. The ripples of this seminal event extended beyond France to England, the United States, and other outlying cultural centers. This time, the movement reverberated beyond the art community and the wealthy avant-garde. Previously, the emergence of a new style remained within the orbit of art magazines and the cognoscenti. Soon after the exhibition ended, American newspapers were touting the new look, followed closely by museums. A collection of European products toured the country in 1925, opening at New York's Metropolitan Museum of Art. "The collection has been brought to America not to stimulate demand for European products nor to encourage copying of European creations, but to bring about an understanding of this important modern movement in design."

Inspired by French department store display window collections of products in the new style, American merchants had window dressers create rooms or tableaus of the new products. Macy's, B. Altman, Lord and Taylor, and Wanamaker's did not expect these dazzlingly modern products to fly out the doors, but they knew the displays would bring in customers. Many of the department stores did not trust their in-house staff to create the atmosphere of modernity that such products demanded. Where else to turn but to specialists in artifice? They hired theatrical designers and promotional designers. Norman Bel Geddes, Henry Dreyfuss, and tableware designer Russel Wright, window dressers all, transitioned from their initial careers into the world of industrial design.

The new direction for designers drew fire from some ranks. James Rennie, in the *New York Times*, criticized Bel Geddes for abandoning theatrical design. Bel Geddes responded in a letter to the paper in a way true to his character—with a manifesto: "The theater is a fickle mistress. We live in an age of industry and business. Industry is the driving force of this age and art in coming generations will have less to do with frames, pedestals, museums, books and concert halls and more to do with people and their life. It is the dominating spirit in this age . . . It is as absurd to condemn an artist of today for applying his ability to industry as it is to condemn Phidias, Giotto or Michelangelo for applying theirs to religion."

The move toward industrial design as a profession led to a skirmish over who exactly was the "father of industrial design." Not surprisingly, Bel Geddes, a man capable of comparing himself to Giotto and

Michelangelo, claimed paternity in 1928, when car manufacturers Robert, Joseph, and Ray Graham and Raymond Paige asked him to design five cars projecting model years from 1929 to 1933. The executives saw the designer's 1933 model and were shocked by the futuristic proposed look. They paid off Bel Geddes and hired two other designers, Amos Northrup and Jules Andrade.

Raymond Loewy, soon to be the profession's most successful free-lance auto designer and never shy about claiming credit, pinpointed his first contract with the Hupp Motor Company as the genesis of the profession. Historian Pulos says no single designer can claim credit for inspiring industrial design:

> Industrial design emerged in the United States as a distinct calling in response to the unique demand of the maturing twentieth-century machine age for individuals who were qualified by intellect, talent and sensitivity to give viable forms to mass-produced objects. The new calling appeared almost simultaneously at many points in order to fill the vacuum left by the inability of craftsmen to anticipate every demand that would be imposed on a manufactured product by an expanding technology and a merchandising commitment to consumer satisfaction. These pioneering industrial designers and other like them were the results of this phenomenon, not its originators.

The early designers were often derided as the field expanded, particularly after the intellectual critics of design came to celebrate the "form follows function" dictates of the modernist movement. The perfection of form sought by the theorists seemingly had to be arrived at through intuition and successive prototypes. The early designers were tackling the design of products that had not existed a decade earlier. The manufacturers cared little for theoretically correct forms. Achievement of a perfect form was all well and good, but if the form didn't sell, then the function didn't really matter. Loewy, Dreyfuss, and Teague looked to achieve, as advertising agencies would establish in the early days of television, a unique selling proposition. Who better to conceive new products than designers who did not subscribe to any theory, save for "give the customers what they want."

Providing consumers with their wants and desires (at least as far as products were concerned) proved to be a lucrative and satisfying pastime for those who had the talent. After his work on Gestetner's duplicators, Loewy began to build his business in earnest. He signed on in 1929 as an art director for a subsidiary of the Westinghouse Corporation, where he designed cabinets for radios. Although he spends the better part of a chapter in *Never Leave Well Enough Alone* denigrating the genre, Loewy produced at least one popular faux architecture design, which was fairly derivative of those by Donald Deskey and Paul Frankl.

The first arena in which designers had overwhelming influence was the automotive industry. Few car companies designed their own bodies for early car models. Typically, outer shells were formed by carriage makers. By 1902, the C. R. Wilson Carriage Company supplied body shells to Oldsmobile, Cadillac, Peerless, and Ford. Ford had sold millions of the Model T, averaging more than one million car sales per year. Ford's nearest competition, Chevrolet, a division of the newly organized General Motors, sold about 330,000 cars per year. Style was the last thing on Henry Ford's mind. He was determined to perfect the assembly line and preferred to give the customer what Henry Ford wanted. General Motors would quickly change that. General Motors grew from the efforts of a Flint, Michigan, carriage merchant named William Crapo Durant. He distrusted the automobile business until he agreed to take over the struggling Buick Motor Company in 1904. Eight years later, Buick sold 8,000 cars, outstripping Ford and Cadillac in annual sales. Durant consolidated thirteen different car companies that same year, forming General Motors.

When Alfred P. Sloan took over General Motors in 1923, he established a plan to introduce customer choice to the car industry. In 1926, Chevrolet offered a vehicle that was available in a variety of colors; not surprisingly, the more colorful line outsold the staid black Fords. Ford retired the Model T and announced a new model, the Model A, instituting annual model changes and color choice. By 1927, General Motors had already founded its styling section and brought in former custom auto body builder Harley Earl to run the shop.

Car design became a proving ground for the four trailblazing designers. Bel Geddes's design for Graham-Paige was celebrated, but never put into production. Teague's design for Marmon Motor Company was put into production as the Marmon 16, one of the more acclaimed cars to presage the streamlined movement in auto design. Walter Dorwin Teague received the commission for the Marmon, but it is generally acknowledged that the car was almost entirely designed by his son W. Dorwin Teague. Just 19 years old at the time, W. Dorwin Teague was a student at the Massachusetts Institute of Technology when he created the Marmon's design. Auto historian Beverly Rae Kimes called the Marmon a "beautifully proportioned, pace-setting design." The car had a sixteen-cylinder engine and was the first major car design to remove much of the surface decoration carmakers used, even the radiator cap.

Bel Geddes's teardrop-shaped vehicle in *Horizons* sported a single vertical fin. The car exists today in model form, but it was never considered for production. Less a catfish than a sort of pregnant guppy, the Bel Geddes car immediately revealed the limitations of automotive teardrop design. One streamlined car that made it into the marketplace was the Dymaxion, a dirigible-like three-wheeled car that also resembled a teardrop. Designed by Buckminster Fuller, who also created a series of circular houses under the Dymaxion brand, the car was underpowered and unstable. It generated a fair amount of favorable publicity until a British journalist was killed when a Dymaxion car collided with another car at the entrance to the 1933 Century of Progress Exhibition in Chicago. Another short-lived "aero" car was the 1935 Scarab, created by aircraft designer William Stout, who also designed the visually arresting Ford Trimotor airliner.

Loewy also was asked to design a car, although he got his first commission through the back seat. Loewy had worked as a consultant for Shelton Looms, designing fabrics and the company offices under the direction of Ely Jacques Kahn. Hupp Motor Company hired Loewy to design upholstery, allowing him to get a foot in the door. Loewy had been dabbling in auto designs well before coming to Hupp. Always cognizant of establishing credit for his ideas, Loewy had filed patents on a headlight, a radiator, and two auto bodies, all blue-sky plans done without the benefit of a client.

Loewy's patent contained some novel features, including a slanted windshield, a streamlined body, a down-slanted hood, and the tucked-under rear end he would use repeatedly in his personal car designs and in some Studebaker models. "The entire car had a dynamic look of motion; the automobiles of the era appeared static."

Hupp Motor Company hired him in the early 1930s. It was founded in 1908 by Robert Craig Hupp, who cut his automotive teeth working for Olds Motor Works and Ford Motor Company before going out on his own. By 1928, Hupp's sales had reached 65,000. But sales plunged after the stock market crash, and the company began looking for a designer who could spark buyers.

According to *Never Leave Well Enough Alone*, Loewy had several offers before going with Hupp. He mentions bonding with Walter Chrysler over locomotive designs. "I was on the verge of being retained by the Chrysler Corporation when a grand fellow named Jack Mitchell of Lennen and Mitchell, the advertising agency, beat him to it." He goes on to claim the moment as no less than the origin of the industrial design profession. "I believe it was the beginning of industrial design as a legitimate profession. For the first time a large corporation accepted the idea of getting outside design advice in the development of their products." It is certain that Loewy's commission was among the first of many entreaties by manufacturers to bring in design consultants, but it was hardly the seminal moment Loewy recalls. He also claimed the company paid him $80,000 a year (almost $1.3 million in today's money), enough to give the fledgling designer some breathing room in his new venture.

Loewy created a pitch to the Hupp production team. As he recalled, "I took the *Detroiter* [train] to Detroit, where a company driver waited for me, and we drove to the plant. Full of enthusiasm I met the Chief Engineer in his office and he started telling me all the things I couldn't do. When it was over, about an hour later, it became apparent that the only things I was allowed to do were to jump out the window, swallow a gallon of enamel, or sit under the ten-ton punch press."

Loewy's tongue-in-cheek narration has a nice ring of resentment as he stumbles into his next meeting with Hupp's production engineer. Never one to shy away from naming names, especially those who wronged him, Loewy hilariously channels James Cagney, not even bothering to name his tormenters: "What he wished me to understand, see, was that whatever I would design that might pass by the Chief Engineer through some oversight, he would catch anyhow." The character Loewy creates for this episode is the international naïf—a well-meaning Frenchman who blithely keeps pitching his services while being ignored by his clients.

The car style that dominated the industry at the time, from the Model A to the Duesenberg, was boxy and vertical. Still influenced by carriage makers, the typical automobile featured square windows and a vertical windscreen, complemented by a long, squared-off hood separated into two pieces. Loewy proposed a car featuring a slanted windshield, rounded windows, and a tapered rear. A sketch in *Never Leave Well Enough Alone* reveals a modern design, particularly the rear, which looks like the Studebaker Avanti. His designs were presented to Hupp's decision makers, and instead of being rejected outright, Loewy learned that his plan would be considered later. "It was a clear case of the polite brushoff; but in my ignorance, I didn't realize it."

Desperate to have his designs considered, he decided to produce a three-dimensional model. "I knew that if they could see the design in the flesh, they would have to admit I was right." Within weeks, paying overtime to a variety of car builders, Loewy had produced a model in 1932 at a personal cost of $18,000 ($316,000 today). The car, a convertible, was a somewhat boxier version of the more streamlined, mass-produced 1934 model, but it served as ample evidence to skeptical Hupp executives. Hupp's decision makers conceded that Loewy's work had merit. The version Loewy presents is one of inspiration, as he decides to spend his own money to "show" the disbelieving manufacturers. His story showcases himself as the ultimate creative, pitching ideas until the client agrees to something, anything. Finally, Loewy can't help but add that while Hupp gutted most of his design ideas, the final product sold well—and that was the important thing.

In a scenario that plays out every model year at every car company in the world, Loewy's new Hupmobile was subtly being changed as the engineering department made running improvements to the design. "What was most objectionable was the fact that it became taller, taller, and taller until it looked like a barn on wheels." Loewy asked an executive to lunch and tried to convince him to lower the car's profile. The executive replied that it made no sense to lower cars because the drivers would feel inferior to travelers sitting above them in streetcars and trucks. Not surprisingly, this argument is not far off from the justifications that twenty-first-century Detroit executives use to rationalize the glut of sports utility vehicles. When the 1932 model was produced, he felt the design had lost much of its dash and glamour, lamenting that it was "too high, too static and blunt-looking." But Loewy admitted there were enough new angles displayed to inspire increased sales.

Although his experience on the 1932 model, the Spyder cabriolet, was frustrating, Hupp executives asked him to design the 1934 model, a four-door sedan. Although Loewy still had trouble transferring his

With Best wishes to the
man who Designed our
Beautiful Car
Dolores and Wally Norman
apr. 1966

The Hupmobile, Raymond Loewy's first automotive design assignment, did not turn out as planned. The company's engineers fought his design ideas, but the final product featured a semi-streamlined body style, integrated headlights, curved bumpers, and other innovations. (Raymond Loewy Archive, Courtesy Hagley Museum and Library)

design vision through the engineering department into the final vehicle, there were considerable breakthroughs, at least for Loewy, in the production model. The auto body was created as a single unit, including the headlights and fenders. The body and its features were rounded wherever possible, including the rooflines, hood, and a slanted windscreen split into three sections. The graceful front bumper, which curved to a point in the center, extended around the front. The rear window, taillights, and front grille were slanted to accentuate the rounded design. The sleek, rounded rear, which featured streamlined fenders integrated into the body, also featured a built-in spare tire in the sheet metal of the trunk, anticipating similar looks used in Lincolns, Cadillacs, and perhaps most memorably in the 1959 Chrysler Imperial (Chrysler aficionados referred to the design feature as "the toilet seat"). Loewy also claimed the integrated spare tire reduced turbulence.

The Hupmobile was Loewy's first jump into waters he had yearned to enter since a childhood filled with sketching planes, trains, and automobiles. The 1934 model shows a definite eye for car design that did not blandly imitate the conventions of Detroit. Still, the Hupmobile was taken off the market in 1936. Hupp Motor Company would not last, closing its doors in 1941 as another casualty of the marketing and manufacturing efficiencies generated by Chrysler, Ford, and General Motors.

For the moment, Loewy had lost his outlet for car design. His experience with Hupp was obviously frustrating, and he vowed to make his next experience with a carmaker much more tolerable. He would get his chance in 1938, with the Studebaker Motor Company, and his design fortunes were about to change for the better. Two major commissions were on the horizon. One would create the impression that design could transform the look of household appliances. The other would provide an introduction into designing the major transportation of the day: trains.

4

Birth of a Salesman
Cold Calls, Clients, and Creativity

When Raymond Loewy decided to concentrate on industrial design, the decision was not rash. He had been marketing himself through his illustrative commissions and accepting different work when the opportunity finally beckoned. Since coming to America in 1919, Loewy had never shied away from expressing his opinions on the look and style of American products. He had not yet made much of a mark in media, but his socializing had paid dividends in regular illustration commissions. A careful look at his autobiography reveals that he had decided on a persona to present to the world—the amusing bon vivant with a talent for design. His book is peppered with scenes where Loewy gives unsolicited opinions on the state of America. "The country was flooded with refrigerators on spindly legs, and others topped by clumsy towering tanks. Typewriters were enormous and sinister looking. Carpet sweepers when stored away took up the greater part of a closet, and telephones looked [this is no pun] disconnected. I felt that the smart manufacturer who would build a well-designed product at a competitive price would have a clear advantage over the rest of the field when things would become tough."

Who better to tell manufacturers what would fly off the shelves than Loewy? He had the cachet of European taste and demeanor and was more than willing to give his opinion on any aesthetic issue. He was emerging within the new profession of industrial design at just the right time. As he put it, "Competition would become fierce, good design could help sales, manufacturers could be convinced—and I was the one to do the job, both the designing and the convincing." Loewy had left France behind, confident that he could put an imprint on his new country, and assumed a more American identity. He would return to France or other parts of Europe countless times and often speak of the intense emotions he felt when returning through New York Harbor. "I felt I belonged here," he wrote in his autobiography. It is perhaps the most well-designed sentence in the book.

After his commissions with Sigmund Gestetner and the Hupp Motor Company, Loewy knew he could make a living as an industrial designer. The question was, could he make a consistent living? He found it easy to communicate the boldness of his opinions and subsequently translate those opinions into salable products—perhaps more than any other designer to emerge in the 1920s.

Loewy's decision to enter the world of industrial design came at a fortuitous time. Electrification of urban areas and the end of mass im- migration, which provided relatively inexpensive labor, spurred the use of machines. Industrial production doubled between 1919 and 1929, and purchasing power rose by almost 20 percent. In 1915, national magazines ran about $25 million in advertisements. By 1923, ad sales totaled $100 million, and by 1929, they had reached $150 million. The ads became the print equivalent of the seductive window displays flanking Main Street, inviting the burgeoning audience to buy more and more merchandise.

Loewy's colleague Henry Dreyfuss described the way manufacturers viewed designers in the early days of the industrial design profession. "Some persons think the industrial designer is the equivalent of a wonder drug like penicillin, to be used when sickness strikes. Actually we are pre- ventative medicine." To sell his services as a product "doctor," Loewy had to hit the road to keep his company afloat. "No one in the manufacturing world had ever heard of industrial design, and no one was interested. My life was a dreary chain of calls on bored listeners."

Characterizing himself as a visiting nuisance that executives couldn't wait to usher out the door, Loewy created an apocryphal dialogue in his autobiography: "Who is that fellow anyway, reeking with a foreign accent, a stranger in the land, to come here in my own office and try to tell me how to run my business? The nerve of some of those Frogs!" Loewy criss- crossed the country making cold calls in nearly every major manufactur- ing city—Akron, Chicago, Cicero, Cleveland, Philadelphia, Pittsburgh,

Toledo, and points west. He writes one of the most vivid passages in *Never Leave Well Enough Alone* about the soul-searing days spent riding trains, then streetcars, and waiting patiently for hours in waiting rooms trying to show a midlevel engineer or executive his sketches and portfolio. In the early years of his career, Loewy's status as a European sophisticate was much more palatable in the metropolitan East than the manufacturing cities of the Midwest.

The marketing of design services required a steady influx of new clients, and in the infancy of the profession, most manufacturers did not walk through the door. Once designers finished a commission, most, if not all, depended on repeat business from pleased clients to maintain their operation. Teague worked year after year for Kodak and National Cash Register Company, while Dreyfuss had long-term agreements with Bell Telephone, Crane Company (a plumbing manufacturer), and John Deere. Loewy also needed to find several "cornerstone" clients to keep his company stable. Most designers gained commissions through face-to-face contact with chief executives. Norman Bel Geddes deigned to meet only with company presidents, an attitude that goes a long way toward explaining his own company's instability.

"And so on and so on, for weeks, for months, for years, with some results once in a while—at ridiculously low fees," Loewy recalled. For two years, he visited Chicago every few months before convincing Sears, Roebuck & Company in 1933 and 1934 to let him design a product. His commission? To redesign the Coldspot refrigerator, one of the company's first attempts to mass-market a major appliance, for $2,500 (about $44,700 today).

Although Loewy credits his dogged cold calls for bringing in the business, in reality he received the Sears commission because another designer turned it down. In 1932, Henry Dreyfuss was the first independent designer hired by Sears, engaged to redesign its wringer washing machine. He enclosed the tub and motor in metal skirting and moved the controls to the top of the machine, calling the model the Toperator. The machine was a huge success. Dreyfuss parlayed that commission into a contract to redesign the General Electric "monitor-top" refrigerator in 1933. When Sears executive Herman Price offered him the contract to redesign their refrigerator, he felt ethically obliged to it turn down and recommended Raymond Loewy to Sears. Although Loewy's persistence as a salesman probably helped seal the commission, the

Raymond Loewy (*right*) and an unidentified designer stand next to the 1934 Sears Coldspot refrigerator. The stark, rounded design was inspired by automobile design. It was Loewy's first nationally popular consumer design. (Raymond Loewy™/® by CMG Worldwide, Inc./www.RaymondLoewy.com)

serendipitous turndown by Dreyfuss certainly counted for as much as the sales calls.

Loewy's new client, perhaps the greatest mass-market merchandiser of the twentieth century, could certainly generate a consistent flow of assignments for the right designer. Sears, Roebuck began as a watch company. Founder Richard Sears, a station agent in North Redwood,

Minnesota, for the Minneapolis and St. Louis Railroad, began his retail career by selling a shipment of watches to other agents on the railroad. He turned a tidy profit. He formed the Sears Watch Company in 1886 in Minneapolis and shortly moved the enterprise to Chicago. He hired watch repairman Alvah Roebuck and expanded the firm into catalog sales. By 1893, the company sold guns, sewing machines, pianos, athletic equipment, and clothing, all through the mail.

Sears's reliance on mail-order and mass merchandising meant that the company depended upon the design whims of its suppliers. Robert E. Wood, a former army officer known to most of his friends and employees as "the General," came to Sears from catalog rival Montgomery Ward, where he had recommended that the business expand into retail stores. Wood had spent decades in the US Army, serving as a World War I logistics officer under Generals Douglas MacArthur and John Pershing, and he recognized the efficiencies and improved shopping experience that retail spaces would provide. Once Sears opened retail outlets, Wood realized that existing department stores appealed largely to women. At Sears stores, where they sought to bring in customers of both genders, the company stocked hardware, auto parts, and "big-ticket" items that were unavailable or more expensive at other retailers. The new lines were effective at keeping men in the aisles. Wood also recognized that appealing products and pleasing or efficient packaging could help market the store's wares.

The company wanted to establish a mass market for appliances where a class market—where only the wealthy could afford to buy—had existed previously. By the start of the machine age, Sears had moved into retail spaces and dialed back its catalog business. By 1931, the company was offering electric refrigerators for $139.50 ($2,200 today). By World War II, all big-ticket items such as refrigerators, washing machines, and appliances were merchandised in retail stores, where customers could personally inspect the larger products. Refrigerator sales rose sevenfold between 1929 and 1935. Radio sales climbed steadily throughout the decade. By 1940, 28 million radios graced American homes, compared to 12 million in 1929.

"When we started our design, the Coldspot unit then on the market was ugly," Loewy wrote. He described the appliance as being ill proportioned and decorated with a "maze" of moldings. Sales of home refrigerators were slowly increasing in the 1930s, but, as design writer Adrian

Forty wrote, "many models with wood casings did not sell the hygienic message."

The design process that Loewy used to create the Coldspot redesign closely mirrored the methods he used in the Gestetner project. Loewy soon hired secondary designers to execute the design prototypes. Loewy and an assistant built a wooden form that was slightly smaller than the original refrigerator. Using clay, the team shaped the piece onto the form, painted it, and finished the look with simulated handles and hardware coated with copper paint and then chrome plating. The full-size model when covered in clay was incredibly heavy, requiring the designer to put the model on bases equipped with rollers. He moved the main motor and pump, which most companies had placed on top to facilitate heat loss, below the appliance, giving the refrigerator large, smooth surfaces with sensuously curved corners and rolled edges. In his autobiography, Loewy spends an inordinate amount of time establishing his insight as an innovator for using clay to create the design, but in fact he had used the method on the Gestetner project, and the auto industry had been using clay models for years to design cars. In *Never Leave Well Enough Alone*, Loewy creates his own history through anecdote. By claiming to be the first to model in clay, the first to research products, the first to design for gas mileage—messages delivered in witty stories rather than lengthy lectures on aesthetics—Loewy gains journalistic legitimacy because the anecdotes are entertaining and easy to recall. The quotable quality of the stories practically ensured that reporters researching a profile would use the anecdotes again and again.

Whatever Loewy's truth regarding his design process, the design for the Coldspot was instantly modern, featuring white surfaces that could be cleaned with a swipe of a cloth. "So seamless [that] a spot called for instant removal," wrote Forty. The Coldspot delivered a message of absolute cleanliness with rounded corners and few dirt-catching crevices. Taking a lesson from his Hupmobile commission, Loewy replaced the steel wire shelving, which was susceptible to rust, with perforated aluminum. He had experimented with aluminum for the Hupmobile's grille, and felt the metal could be shaped into a pleasing look much more easily than steel. In an innovation that could be revived today, Loewy created a long, full-length vertical bar latch. The latch was designed to open the door by hand, or with an elbow if a housewife had both hands occupied with bags or packages. Loewy also designed a remote-control latch that would open the door using a foot-operated pedal. "He designed a product that people felt good about," wrote design critic Thomas Hine. "He packaged a dirty, noisy thing into an object of desire."

The Coldspot commission was one of the first industrial design assignments to receive large media exposure through trade journals and the business press. The advancing popularity of business magazines and lifestyle magazines such as *House Beautiful* provided a springboard for the new products churned out by manufacturers and retailers, and, in turn, to the professionals who designed those products. Loewy found that professional organizations and a few colleges were interested in lectures on the Coldspot case study and accepted offers to make presentations on the project. "Form followed function if the function of a product is to be sold," Hine wrote.

The interior of the Coldspot refrigerator used technology taken from Loewy's work in the auto industry, such as creating shelf racks from extruded aluminum and designing crisper drawers to have streamlined "speed lines." (Raymond Loewy Archive, Courtesy Hagley Museum and Library)

Sales of the new refrigerator grew from 15,000 to 275,000 in five years. Loewy was commissioned to design the 1935 model for $7,500 (about $131,000 today) and received $25,000 ($436,000 today) when sales had reached a peak. Part of those sales numbers is attributable to Sears's decision to make the Coldspot 6 cubic feet in size in a market where 4 cubic feet had been the standard. In fact, an early history of the Sears empire attributed the uptick in sales almost exclusively to the larger size differential. Sears found that the larger appliance cost just $5 more to manufacture. Its catalog featured a description of the design prominently in its advertising copy. The redesign of the Dreyfuss washing machine and the Loewy Coldspot refrigerator gave Sears a major share in home appliance sales, a presence it relinquished eight decades later. Loewy's success launched his company, allowing him to hire his first permanent design employee, Robert Jordan Harper.

Changes in the Sears Coldspot designs for the next few years kept Loewy's company busy. A case history written by the Loewy office said that the original Coldspot design was "a step in the evolution toward perfection." The subsequent designs made significant—if minute— changes. The 1936 model featured a chrome strip down the middle of the door, as well as a nameplate above the latch. Loewy added a storage compartment into the base. The 1937 model featured more changes. They eliminated the chrome strip, replaced by a subtle V-shaped peak, giving the refrigerator door the look of an automotive hood. The 1938 version reintroduced a gap in front supported by front legs wrapped with chrome tubing. Loewy extended the sides of the appliance to floor level, streamlining the entire box and substituting a handle on the door for the release bar. Soon after the last redesign, which revealed Loewy as being amenable to yearly model changes that mirrored the "planned obsolescence" of the car industry, he and Sears parted ways. He landed the contract to design refrigerators for Frigidaire, a subsidiary of General Motors, in 1939, a commission he would maintain for decades.

The selling of industrial design to corporate clients was difficult at first, but some designers were better at it than others. One of the first to standardize a system for publicity, Norman Bel Geddes included the sales procedure in an employee manual. Before the firm made a sales pitch to a company, Bel Geddes's public relations department would submit an article to a trade journal on a general topic that happened to address the target company's business. If the target replied with an inquiry, the PR department would send a stack of press clippings extolling Bel Geddes's work. Finally, if a contract was secured, the media was inundated with press releases about the firm's stellar work for its client.

The design profession was wide open for anyone willing to call themselves a designer, thanks in no small part to the publicity generated by Loewy, Bel Geddes, Teague, and Dreyfuss. Yet Loewy takes pains to point out in his memoir that only a chosen few were deserving of the title. "It was a stampede," he wrote. "Some of them, having persuasive ability, managed to get design commissions out of credulous manufacturers and they produced an unbelievable amount of imbecilic stuff."

Singling out Egmont Arens, Henry Dreyfuss, Ray Patten, Walter Dorwin Teague, Harold Van Doren, Russel Wright, and others as worthy designers, Loewy castigates the rest. "We had to contend with a group of twenty or thirty crackpot commercial artists, decorators, etc. without experience, taste, talent or integrity, who called themselves industrial designers. They nearly killed in the nest the young profession that we were trying to nurture through adolescence." Most of the people Loewy was disparaging were designers in specific industries rather than the multiclient design firms pioneered by Loewy and his peers. Loewy recognized quickly that the most important product he had to sell was Raymond Loewy. He became the chief salesman as well as the final arbiter and hard-to-please editor of the firm's designs. An office expansion, a move made in 1933 as the country was mired in the economic downturn of the Great Depression, must have seemed risky, even to the supremely self-confident Loewy, yet he wrote of the experience as if he were considering buying a new suit.

"So I took an office on the fifty-fourth floor of a skyscraper at 500 Fifth Avenue. I engaged two good designers, a secretary, and we got under way. It was a success from the start." The key phrase in his recollection is "engaged two designers," neither of them identified but clearly expected to shoulder the design work. The business thrived, and soon Loewy, snowed under by paperwork, hired a business manager expected to do double duty as manager and secondary rainmaker for the new business. Loewy writes he hired a hail-fellow-well-met outgoing salesman, a former athlete with society connections. As he tells it, the new man prepped for his first presentation by downing several double martinis at the New York restaurant 21. The meeting nearly escalated into fisticuffs when Loewy's representative came on too strong to a corporate tycoon.

By setting up the classic American go-getter stereotype in the story, the author sets the stage for an alternative, his own persona as the urbane aesthete. So, Loewy sacked the would-be manager and decided to become the public face of his enterprise. He would oversee the designs, but most of the work was to be done by others. His personality was such that public sales might be the last thing he was suited for. He was always "Mr. Loewy" at the office, never "Raymond." Even his autobiography reveals little about himself or his personal life, devoting just one paragraph to his marriage to Jean Thomson. "She spoke fluent French, we soon felt an irresistible attraction for each other, and we married in 1931." Just one sentence of detail, and then Loewy meanders through a story about the couple's shaggy dog.

After the Sears account increased the designer's public profile, he found he had more work than he could easily handle. Loewy chose business over aesthetics, delegating day-to-day designing in favor of becoming a benevolent despot, content to oversee design aesthetics, saving a few choice assignments to oversee himself. His decision to expand during the Great Depression is a textbook illustration of savvy management (successful corporations spend more on research and development during downturns). His first expansion and reorganization included promoting three designers to executive positions, hiring a business manager, and creating a publicity and public relations unit.

The expansion garnered immediate dividends. The company's business expanded, and two years later, Loewy added a staff of technicians and engineers, a model shop, and a clay- and plaster-modeling department. As the firm prospered, Loewy came up with "a fundamental rule," what would today be described as a mission statement: "Nothing is to come out of R. L. offices until it has been checked and double-checked for practicability and manufacturability. Heads of divisions will be held directly responsible for the observance of this policy."

There was a business mission, but to hear Loewy tell it, the design office was an amalgam of intellectual salon, assembly line, and cocktail party. Throughout his autobiography, Loewy talks of visitors wading through canapés, water coolers filled with goldfish, and, on one occasion, a performance by an unknown singer from Hoboken, New Jersey, by the name of Frank Sinatra. It's hard to imagine that the young bigband singer would have played an office party, but the 1951 autobiography did not inspire a response from the litigious singer. One of the ways Loewy used the office centered on what today's businesspeople would call networking. He specifically credits a network of contacts with helping his early business take off. Perhaps the best known of these early supporters were Charles and Stuart Symington.

Stuart Symington is the better-known name today, but both Symingtons were influential in New York society and in business. Secretary of the air force under John F. Kennedy and a former US senator representing Missouri from 1953 to 1976, Stuart Symington met Loewy in 1936, and the two remained lifelong friends. When Symington was named the first secretary of the air force in 1947, Loewy designed his office. Symington also arranged for a Japanese design tour for Raymond and his second wife, Viola, following Loewy's retirement. Loewy made inroads into corporate commissions by an introductory letter written by Charles Symington, an executive for Colonial Radio Company. Colonial was not one of Loewy's more renowned clients. Loewy's best-known design for them was a 1933 radio built to resemble a desk globe. While working for Colonial, Loewy had shown Charles Symington, with whom Loewy would soon become socially friendly, several sketches of streamlined locomotives. Less than a month later, the designer met with Martin W. Clement, a friend of Symington's and president of the Pennsylvania Railroad. Loewy carried with him the same sketches he had shown Symington.

Clement was president of the largest railroad in the world, a position it had held for most of its history. It owned 10,000 miles of track and moved more passengers and freight than any other railway on the planet. By 1924, it owned 7,000 locomotives and 282,000 railcars. The Pennsy system stretched from New York to Chicago.

Loewy arrived in Philadelphia for his meeting with Clement. He had made an appointment with the rail executive on the strength of Symington's letter. "Well, young man, what can you do for this railroad?" Clement asked Loewy. The two men bonded when Clement asked the young designer how he had spent the war years. Clement had served on the French front, and Loewy wrote ironically, "Apparently, he kept a pleasant memory of the experience." Loewy goes on to re-create the conversation, certainly one of the most important of his life:

"Have you ever designed any railroad equipment?" [Clement] said in his stentorian voice.

"No, but I have been dreaming about it for the past 20 years."

"So you like being in America?"

"I couldn't live away from it."

Clement tried what Loewy called ruefully, "the brush off."

"Well, all right. We don't have anything for you now, but we'll get in touch with you if something turns up."

Loewy gambled.

"Mr. Clement, I would like to start right now. I know I can do valuable things for you and the railroad and I would like a chance to prove it to you. Can't you find one single design problem to give me now, today?"

"What did you have in mind?"

"A locomotive."

Loewy describes spending an uncomfortable and unbearably tense few minutes before Clement summoned a secretary and asked for a junior executive, a Mr. Brown, to come over.

"This Frenchman has been recommended to me by Symington. He has a good war record and he likes it here. He thinks he can be a designer. He insists on doing a design job for us so he can prove how good he is. Look here, Brown, the trashcans in the New York terminal are terrible. Why don't we have better ones? I want you to make arrangements for this Frenchman to have a look at our trashcans. Maybe he'll be able to do something about it."

Clement addressed Loewy, "Go to it."

Loewy went to it by staking out Pennsylvania Station, then the architectural jewel of American railway stations. He spent three days "looking over the trash can situation." He designed three prototypes, and several models were built by staff designers, although Clement did not sign off on the design. At this point, Loewy was likely still working on few accounts, as he had time not only to observe how station trash cans were used, but also to spend several days watching his design models being used. Loewy returned to New York only to be summoned back to the president's office in Philadelphia. Clement greeted him with, "How's the great trashcan specialist today?" Loewy, anxious to see how his design was received, sat through Clement's small talk about French cooking, the Great War, and French culture and finally asked what Clement thought of the design. Clement replied, "Young man, in this railroad, we never discuss problems that have been solved." Pennsylvania Railroad records showed Loewy and company were paid $119.31 (just over $2,060 today) in 1936 for the design. The trash cans were originally installed in Philadelphia's 30th Street Station and later were used in New York and Pittsburgh as well.

Clement's next move was to call in F. W. Hankins, the Pennsylvania Railroad's chief of motive power. Hankins brought with him sketches for the railroad's new electric locomotive, the GG-1. After a childhood spent looking at and sketching trains, railcars, and other railway equipment, Raymond Loewy was about to work on a locomotive.

5

Big Engines
Emergence of a Design Genius

On a blustery, chilly autumn day in 1939, a dapper-looking man in his mid-40s, wearing a black overcoat complemented by a scarf and slouch hat, climbed onto a railway station platform in Fort Wayne, Indiana, to watch the approach of a train. Amid billowing steam, shrouds of smoke, and a rumbling noise loud enough to make him tremble, the man watched an enormous locomotive fly past. The sleek, sculpted machine, effortlessly pulling a passenger train, resembled a speeding bullet on wheels.

He later wrote in *Industrial Design*:

On a straight stretch of track without any curves for miles; I waited for the [engine] to pass through at full speed. I stood on the platform and saw it coming from a distance at 120 miles per hour. It flashed by like a steel thunderbolt, the ground shaking under me, in a blast of air that almost sucked me into its whirlwind. Approximately a million pounds of locomotive were crashing through near me. I felt shaken and overwhelmed by an unforgettable feeling of power, by a sense of pride at the sight of what I had helped create in a quick sketch six inches wide on a scrap of paper. For the first time, perhaps, I realized that I had, after all, contributed something to a great nation that had taken me in and that I loved so deeply.

Raymond Loewy stood on the Indiana platform that day watching the first locomotive he designed from sketchpad to rolling stock, the streamlined S-1 steam engine. It was built by the Pennsylvania Railroad to haul its most popular passenger train, the *Broadway Limited*, from Chicago to New York. The largest ever, the locomotive measured 140 feet long and weighed more than 500 tons. When it debuted in 1939, it was capable of hauling a multicar train at speeds up to 140 miles per hour. Although other companies and designers had produced artfully designed locomotives, the S-1 is the engine most commonly associated with the term "streamlined."

Aerodynamic efficiency was not new. George Stephenson and Nicholas Wood, two British railroad engineers, investigated rail engines for optimum shapes in 1818. In 1834, a British count performed aerodynamic tests on the Liverpool and Manchester Railway. In 1865, a Roxbury, Massachusetts, minister named Samuel Calthrop patented a design for a "wind-splitting" train (which looked remarkably like today's Amtrak's *Acela* commuter train), the American version of the high-speed train. Calthrop's smooth, tapered locomotive and conical nose transitioned to articulated passenger cars and a tapering tail caboose. The Union Pacific Railroad brought out a knife-shaped railcar in 1908, designed by its supervisor of motive power William McKeen. The McKeen Motor Car came to a point in the front, like a ship's prow.

By the time of the Loewy commission, the Pennsylvania Railroad had been lagging in the streamlining wars. The rival railroads all had altruistic reasons for streamlined design. Sleeker vehicles saved fuel, and some designs reduced the soot and ash that plagued travelers. But the railways of the 1930s were losing customers to other transportation alternatives, so the industry took a page from the automotive playbook and introduced trains that not only functioned as machines but also inspired excited passengers by becoming rolling sculptures. The success of the Pennsy—which started in 1849 with two locomotives, two passenger cars, a baggage car, and 69 miles of track in Pennsylvania between Harrisburg and Lewistown—bred complacency. By the turn of the century, led until the 1870s by visionary executive J. Edgar Thomson, the line had been extended to Philadelphia and Pittsburgh, and additional lines stretched even farther, to St. Louis, Chicago, and Cincinnati. The railroad expanded again after 1900, led by company president Alexander Cassatt, after purchasing the Long Island Railroad and major blocks of stock in the Chesapeake & Ohio, Norfolk Southern, Baltimore & Ohio, and Reading Railroads. Cassatt culminated his acquisitions by extending the railroad into New York City and opening Pennsylvania Station. These successes gave Pennsylvania Railroad executives, and most rail executives in general, a false sense of superiority. As the country entered the automobile age, railroads were ill prepared to face competition.

Streamliners not only brought more riders back into railcars, but also saved money for most of the railroads who introduced them. Railroad engineers estimated that air drag sucked away three-quarters of the power generated by a nonstreamlined train. In 1939, railroads experienced gradual declines in customer loyalty as bus lines began taking passengers across country, automobiles helped Americans travel much more independently than a decade earlier, and the development of early highways made traveling easier. To counteract these trends, railroad companies sought to bring riders back to their trains by marketing plush, attractively designed passenger cars pulled by machines that looked fast just standing still—these were rolling visual fantasies executed in steel.

After Loewy's wastebasket commission, the designer was looking to make a lasting impression on the country's rail system. Although Loewy's creation story for his Pennsy work centers on his triumph over yet another set of dour, disbelieving executives, the fact is that the large corporation had often looked to outside contractors for design help. Its cars and locomotives had been designed in many cases by Pullman, American Car and Foundry, and Baldwin Locomotive Works. The railroad's most memorable stations had been created by some of the leading architects of the era, including the firm McKim, Mead and White (New York's Pennsylvania Station), Daniel Burnham (Pittsburgh's depot and Washington, DC's Union Station), and Frank Furness. Architect Paul Philippe Cret, who headed the University of Pennsylvania School of Architecture (and who would later collaborate with Loewy on interior train designs), was hired to create Pennsy designs for the architectural treatment of bridges.

The diesel railroad locomotive offered many advantages over steam-powered engines. They ran cleaner, required relatively less maintenance, and did not need an elaborate infrastructure of coal stations and water towers to keep them going. Diesel locomotives also had no overhead expenses for coal hauling and ash removal. Train crews for steamers had to be on hand to stoke fires in idle steam engines to keep the boiler pressurized.

Electric trains were the most efficient and relatively easy to maintain. Their drawback was infrastructure—the idea of stringing electric lines across the country's railroad tracks was too immense a job to contemplate. Electric locomotives were used only for smaller rail lines in areas where steam power was deemed inconvenient. For example, the Long Island Railroad was electrified in 1905, shortly after New York City banned the use of steam locomotives within its city limits to reduce pollution. By 1933, the Pennsylvania Railroad ran electric engines from New York to both Philadelphia and Washington, DC.

Diesel power was quickly adapted for railroad use after the original diesel engine was patented in 1892. The power generated by the oil-burning engine did not drive the locomotive. Instead, it generated electricity to drive axle-mounted electric motors. The first diesel engines were massive and heavy, but as they became more efficient, it became clear that diesel locomotives were a workable alternative to steam power. In 1930, General Electric developed a lightweight engine that generated enough power to drive a small train—the *Burlington Zephyr*. It ran a nonstop route from Denver to Chicago, a trip of 1,015 miles, in 13 hours.

For railroads, the initial investment in a diesel engine was substantial, more than double the cost of a steam engine. But the high initial cost was offset by reduced fuel and maintenance costs (by World War II, diesel fuel costs were half of that required for steam). Steam engines also were designed by railroads to fit the requirements of a particular route. When the engine's useful life was over, steamers were cut apart for scrap. Diesel engines, which had a standardized design, could be used in any situation and could be resold to smaller rail lines to recoup some of the investment cost.

The shape of a diesel engine—essentially a rectangular box—allowed designers to place the cab and driver's compartment of the engine in front, making it easier to create attractive, cutting-edge engine designs. The absence of a rising smokestack or visible outside cylinders made a diesel exterior more visually pleasing. Such streamlined diesels as the Burlington Railroad's *Zephyr* and the Union Pacific Railroad's *City of Salina* demonstrated to competing railroads that the more attractive a locomotive was, the more passengers it would attract.

Loewy's first attempt at locomotive design came much earlier than the streamlined S-1. He had suitably impressed Pennsylvania Railroad President Martin W. Clement with his trash can—enough to merit a larger commission—which turned out to be the railroad's new electric locomotive, the GG-1. In both of his books, Loewy regales the reader with his cosmetic changes to the electric engine. He also artfully omits the fact that, mechanically, much of the locomotive was designed before he laid eyes on it. Still, Loewy knew he had the commission of a lifetime. The Pennsylvania Railroad was recognized as the largest rail company in the world, with more than 100,000 employees and serving almost half of the citizens of the United States. Perhaps more than any other railroad, the company built much of its own rolling stock and infrastructure.

Loewy's first major assignment for the railroad, the GG-1, was an existing project that inspired Clement to ask whether Loewy perceived any flaws. The original design, which featured a distinctive central cab and sloping "fenders," has long been credited to Loewy, but the first design for the GG-1 more likely emerged from the drawing table of a Pittsburgh-based industrial designer, Donald Dohner. A designer-educator, Dohner started the first academic industrial design program in the United States at the Carnegie Institute of Technology (now Carnegie Mellon University) in 1934 and moved to the Pratt Institute in 1935, where he established the school's industrial design program. Before entering higher education, he worked for Westinghouse from 1930 to 1934 as head of the Art in Engineering Department.

Dohner's design epiphany for the first GG-1 prototypes was the central "visibility cab," which allowed engineers to more clearly see the tracks in either forward or in reverse. He first used the concept in a series of blocky Westinghouse switcher engines. Dohner's work on the GG-1 came when the Pennsylvania Railroad asked General Electric, Westinghouse, Baldwin Locomotive Works, and independent engineer George Gibbs to create a high-speed electric locomotive. The project started in 1934. An early decision to improve safety for the crews resulted in a centrally positioned "steeple cab." Dohner's visibility engines were clear antecedents to the GG-1, his early models suggesting the basic shape and design of what would become the Loewy version. The first prototype version of the GG-1, called "Rivets" by Pennsy railroaders, incorporated some aspects of the Dohner models, but its appearance was bulky and literally riveted together.

The Pennsylvania Railroad hired Loewy to oversee the design of the GG-1 into mass production in 1934. Dohner subsequently left the GG-1 project to enter academia. At the time, a case could have been made that Dohner was the more celebrated designer, although both Loewy and Dohner were listed as being among the top ten industrial designers in the United States in the February 1934 issue of *Fortune* magazine. When Pennsylvania Railroad executive Clement asked Loewy his opinion of the "Rivets" prototype, Loewy, according to his account, was reluctant to speak his mind but instead offered, "It looks powerful and rugged, yet I believe it can be further improved."

If the engine could be improved, the same goes for Loewy's tale of design. In his subsequent retellings of the Pennsylvania Railroad commissions, he introduces a story he uses repeatedly in different permutations. Asking the two executives if he could take the sketches of the engine back to his office, Loewy had conceived an improvement plan by the time his train had pulled into Penn Station. The trip, which would have taken an hour or two, inspired the designer in this way: "I was thinking in terms of simplification. I wanted to show these men that I was no long-hair artist trying to pretty up a 6,000-horsepower locomotive." Loewy decided to recommend butt-welding the train's outer shell and lowering it onto the chassis, a technique often seen in automotive manufacturing. The butt welds would eliminate thousands of rivets protruding from the shell, making the surface smoother and easier to clean and maintain. He presented an airbrushed rendering of the new design (done by one of his employees), and Clement and Fred Hankins, the railroad's chief of motive power, were taken aback by the concept.

Loewy returned to Philadelphia several times to inquire whether the engine was to be built, but he received no answer. He dropped in on the railroad's engineering department and discovered a draftsman drawing his locomotive. Hankins happened to be in the office and came forward to say that the company was going to build a full-size mockup of the design. He invited Loewy to come to Wilmington, Delaware, to see the model.

Hankins had previously written in a memo to another executive that the railroad should hire "someone with good taste and someone who is generally acquainted with passenger design from the viewpoint of what has been done abroad, as well as in this country. However, I am opposed to the employment or engaging of these so-called professional designers

Loewy and several unidentified Pennsylvania Railroad workers next to the workhorse electric locomotive, the GG-1. Loewy was hired after the locomotive was designed by Donald Dohner. He created the smooth, butt-welded outer shell and added speed lines to the locomotive's sleek look. Many GG-1s are still in museum collections. (Raymond Loewy™/® by CMG Worldwide, Inc. / www.RaymondLoewy.com)

who have been brought up principally in plants where wall-paper designs are developed." The ideal candidate "will have to be a pretty high type man and must be something over and above a wall-paper artist." Loewy, whose familiarity with wallpaper designs reached back to World War I trenches, fit most of Hankins's criteria.

Loewy's first look at the life-size prototype inspired a few corrections and additions to the original design. He also added cosmetic touches, such as five painted "speed lines" along the upper length of the shell to suggest motion. As engineers and rail workers assembled the prototype for the GG-1, Loewy climbed a ladder and sketched out where the lines should go. He would later claim that the speed lines, called "cat whiskers" by railroaders, were a safety precaution to help track workers see the giant electric locomotive, which made little noise compared to its steam-powered relatives. In reality, they didn't much help with visibility, but they accentuated the down-sweeping form of the engine. The engine retained the central cab and had a wheel layout that allowed it to operate in either direction. The first 1934 test run for the locomotive was filled with dignitaries, including Cordell Hull, secretary of state, Harold Ickes, head of the Public Works Administration, and the governor of Pennsylvania, Gifford Pinchot. The Pennsylvania Railroad went on to build more than 130 of these locomotives. Owing to its durability, it is the most commonly seen Loewy locomotive today, displayed in many railroad museums across the country.

After the success of the GG-1, Loewy and his firm signed an exclusive Pennsylvania Railroad contract for an annual fee of $20,000 (around $360,000 today). The retainer later rose to $25,000 (just over $431,000 today) and that, coupled with labor and overtime charges, earned Loewy's company more than $100,000 each year (which, by 1939, would be worth $1.7 million today) through the 1930s.

"One thing led to another until we were called in almost daily for consultation on some new problem," Loewy wrote of his relationship with the company. "It could be anything: the color of a ferryboat, the design of a menu, a new signal tower, a bridge over the Potomac, a coffee cup, or the design of a bronze tablet for a retiring executive."

The Pennsylvania Railroad constantly haggled over billable hours and was often at odds with Loewy and Associates' cavalier attitude toward accountability. Loewy's designers often would expand an assignment beyond what was asked for, and every now and then presented a design that hadn't been commissioned. Glenn Porter, curator emeritus for the Hagley Museum, which houses an extensive Loewy archive, discovered a note from the Pennsy's chief of motive power asking Loewy to explain a proposal to change the clocks in Pennsylvania Station. Loewy replied, "We received no authorization from the Pennsylvania Railroad to do this work, but it started because so many people called our attention to the fact that the complicated design of the numerals, lettering and hands make it very difficult to tell the time at a quick glance. The attached photograph will illustrate the point." The railroad got the point, but Loewy's firm never received a check. Still, the railroad bought plenty of other Loewy suggestions.

Although the GG-1 was a success, the company was overwhelmingly invested in steam-powered engines, and it decided to double down on its investment. The S-1 was known to all those who worked on or around it as "the Big Engine." It was arguably Loewy's biggest commission, both in size and in terms of business success. Loewy was photographed more with this design than that of almost any other product. He often posed perched on the "cow catcher" or posed just below the rail on the massive torpedo-like boiler.

Some American railroads used their own engineers and draftsmen to design locomotives, much like American automobile companies still do today. But as the need for dazzling designs became more urgent, rail lines turned to outside help.

In 1927, the New York Central hired a German painter, Otto Kuhler (1894–1976), who had settled in Pittsburgh three years earlier, to design its J-1 Hudson steam locomotive. He created a series of paintings for the design. The engine was never built, but Kuhler was subsequently hired as a designer for the American Locomotive Company. He also went on to create several conversions of existing locomotives, including the Lehigh Valley Railroad's *Black Diamond* and *John Wilkes*, as well as the Baltimore and Ohio engine *Royal Blue*, the first streamliner put into use. Kuhler had received engineering training as a young man working at his family's steelworks. He left to work as an automotive body designer in Germany and Belgium, and during World War I helped design a railroad in Belgium. He emigrated to Pittsburgh in 1923 and started an art studio providing sketches and paintings of industrial scenes. Kuhler's designs were pleasing, but they were not trend setting. In fact, most streamliners differed only in the aesthetic detailing. The designer who might have whetted the appetites of the general public for leading-edge futuristic engines, Norman Bel Geddes, showcased several streamlined train designs in his book *Horizons*. No rail line ever built them.

The Pennsylvania Railroad was heavily invested in steam engines, as was the company's main competitor, the New York Central Railroad. But in the early 1930s, most steam engines looked much the same as they did in 1900—huge and hulking, studded with gears, rivets, and grime. The Pennsylvania Railroad wanted to reinvent itself for the modern world by redefining the image of the steam locomotive. The engineers of the major railways also knew, based on a series of 1931 experiments performed by Westinghouse engineer Oscar Tietjens, that streamlining could save up to third of its expended power, in turn increasing profits.

In the 1930s, streamlining became more popular as several designers, including Bel Geddes and Loewy, designed sketches of propeller-driven streamlined railcars, inspired by air travel. The first truly streamlined train put into production was the *City of Salina*, built by the Pullman Car and Manufacturing Company for the Union Pacific Railroad. Its interior, using a pale blue-and-white color palette, was designed by Marie Harriman, wife of Union Pacific president Averill Harriman. The exterior design was conceived by engineers within the company, who were influenced by several of the experimental engines drawn by Bel Geddes in *Horizons*. Delivered in 1933, the *City of Salina* engine resembled the head of a snake, but the 204-foot train's 110-mile-per-hour speed put it at the forefront of streamliners. The engine attracted hordes of visitors at the 1934 Century of Progress Exhibition in Chicago. The exterior featured a brown paint job along the train skirting and above the windows. The side panels were canary yellow. The engine featured a yawning air scoop that flowed into a four-window cab with a cowled headlight finishing the serpentine design.

Many early streamliner diesel designs used the down-slanted and rounded front profile of the Burlington Railroad's *Zephyr* (named for the Greek god of the west wind), manufactured and designed by E. G. Budd Manufacturing. The Burlington train, which was soon renamed the *Pioneer Zephyr*, soon was joined by two other Zephyr trains, the *Twin Zephyr* (serving Minneapolis) and the *Denver Zephyr*. The interiors of all three trains were designed by Paul Cret, the architect who would soon collaborate with Raymond Loewy on the *Broadway Limited*. The exterior's bright colors were set off by chromium or stainless steel. The Zephyr look, called a "shroud" design, was used on the first streamlined steam locomotives, the New York Central's *Commodore Vanderbilt* (1934), as well as the diesel-powered Milwaukee Road's *Hiawatha* (1935) and Gulf, Mobile and Ohio's *Rebel*, both designed by Otto Kuhler.

Henry Dreyfuss also designed a shrouded steam engine, the New York Central's *Mercury*, which sported driving wheels that were painted white and illuminated at night by hidden spotlights. The design essentially draped a cowling over a Hudson locomotive, giving the engine a sleek look of an "inverted bathtub." Dreyfuss was the first designer to be given carte blanche to design not only the locomotive, but also the entire train. He redesigned existing New York Central railcars and lowered window sills to give passengers panoramic views. The "shroud" designs became ubiquitous, and many rail lines refitted steam locomotives with these streamlined shrouds, often designed by in-house engineers.

Despite its reputation as a forward-thinking company, the Pennsy was rather late to embrace streamlining. The railroad had huge engine-building shops in Altoona and Lima, Ohio, as well as access to the Baldwin Locomotive Works outside of Philadelphia. Its engine designs were done by in-house engineers, all individuals capable of designing a mechanically superior locomotive. Most of these engineers saw no reason to change the basic look of a steam locomotive, which had remained unchanged since the 1860s. But the Pennsylvania Railroad did not employ anyone capable of finding poetry and the sleek suggestion of motion within the tons of iron and steel needed to create a locomotive.

For that they needed Raymond Loewy.

But first, after Loewy completed the GG-1 assignment, the railroad asked the designer in 1936 to give the Pennsy's workhorse engine, the K-4 Pacific, a streamlined profile. The redesign essentially allowed Loewy to work out some theories for his later work on the fully conceived S-1. The makeover transformed the engine's front end into a bullet-nosed projectile and extended skirting to hide the engine's moving parts. The new look, called "the Torpedo" by railroaders, included multiple access doors on the outer cladding that allowed mechanics to access engine parts and car couplers.

Loewy asked Fred Hankins if he could observe a ride in a K-4 at high speed. He showed up for the test dressed for the experience in a sweater, anorak, heavy gloves, tweed cap, and goggles. Riding the stretch of track between Chicago, Illinois, and Fort Wayne, Indiana, Loewy took a length of wood, tied a white ribbon to it, and lifted it into the locomotive's airstream. He soon saw a pattern emerge, which was used to help design the smokestack. "I discovered how to catch a cold in a few minutes," he wrote. He also discovered that there were no toilets in most locomotives and made a note to design one into his proposal.

The final engine design was tested in clay model form in the wind tunnel at the Guggenheim Aerodynamic Laboratory of New York University. The engineers calculated that the streamlining would conserve as much as 300 horsepower when operating at 90 to 100 miles per hour. Loewy also incorporated a rubber diaphragm that closed the gap between the locomotive and the coal tender, giving the head of the train a long, smooth profile. According to rail historian Robert Reed, the tests were almost irrelevant, particularly to Pennsy passengers. Patrons did not really care if the train was fast; what mattered to them was that the train *looked* fast. One of the most annoying aspects of steam locomotive travel was locomotive smoke and cinders wafting into passenger cars. Loewy's solution to the problem fitted a smoke-lifting device for the K-4 that channeled the smoke upward from a stack recessed into the engine cowling. The emphasis on looking fast was completed by Loewy's use of striping, complemented by stainless steel handrails that stretched around the boiler and front platform.

Similarly, the final design for the S-1 in 1939 accentuated the immense cylindrical boiler that dominated the front of all steam engines. Whereas diesel design accentuated the front cab using a snakelike shroud or wedge form, steam streamliners used a "bullet" or "torpedo" shape. Loewy's chief rival, Henry Dreyfuss, used a torpedo-inspired design on his *20th Century Limited* engine for the New York Central. His boiler concept rounded into a perfectly cylindrical headlight, which was bisected by a large fin that gave the engine the profile of a Trojan helmet. The *20th Century Limited* was a popular success with passengers, prompting Charles Sheeler, the American precisionist painter, to use the engine for a brilliantly conceived *Fortune* magazine cover called *Rolling Power*.

New York Central's *20th Century Limited*, built in 1938, was the first train to be marketed as a travel convenience. It also had a practical purpose, as the railroads were losing passengers who were weary of multistop journeys. The idea of a high-speed train making limited stops between New York and Chicago dated to the 1890s. A New York Central executive, George H. Daniels, realized that speed and exclusivity would appeal to travelers. A special train made its first run to carry New York Central passengers to the 1893 Columbian Exposition, covering 960 miles in 20 hours. The *20th Century Limited* became a regularly scheduled train in 1902, sizzling along the tracks at speeds exceeding 100 miles per hour. New York Central's promotional literature claimed the train did not stop for water to refill boilers. Instead, the locomotive scooped in water from long pans laid between the tracks. Service on the train made passengers feel like royalty. Riders could have maid, valet, or secretarial service, and telephones were on board by 1905. When the train rolled into New York City's Grand Central Terminal, attendants rolled out a red carpet for all disembarking passengers.

The design of the *20th Century Limited* was impressive, and the huge locomotive rivaled Loewy's work for innovation. The best work in Dreyfuss's assignment came in his overall conception for the train interiors. The dining car could be separated into smaller dining rooms, and a closed-circuit phone system allowed riders to call ahead for reservations. All the tableware, linens, and even the sugar cube wraps sported Dreyfuss's innovative logotype, a vertical series of stacked bars called "the chimney" by New York Central fans, that travelers saw repeated on nearly every printable surface from Grand Central Terminal to the major stations on the way to Chicago.

Loewy, however, had his own ideas on how to give style and speed to rolling power. Improving on the streamlined redesign he had done for the existing Pennsy K-4 locomotive, Loewy created the S-1 as a low-slung, torpedo-shaped boiler that flared slightly at the sides. Instead of being made as a perfect cylinder, the boiler was gracefully rounded and flared into a centralized headlight, under which Loewy mounted a stainless-steel handrail. The driving wheels were covered with skirting painted with three gold "speed lines." The huge boiler was balanced visually by a rounded "cowcatcher" platform that jutted out beyond the nose and covered the forward wheel truck. The forward platform was painted with three additional speed lines. Loewy had again tested a clay model of the prototype in the Guggenheim wind tunnel and improved on a smoke deflector design he had tried with his previous K-4 streamline design. For both the S-1 and the K-4 redesign, Loewy claimed he had spent hundreds of hours riding in the cabs of Pennsylvania Railroad locomotives.

The S-1 locomotive was the first "duplex" steam engine, built with four cylinders mounted on a rigid frame. One of the massive engine's primary design problems centered on keeping the twenty wheels from overwhelming the look of the design. Loewy solved the problem by integrating a cowled skirting, forming a stage deck that effectively showcased and framed the long cylindrical boiler. The ornamentation of the engine was deemphasized to better display the immensity of the locomotive. To commuters and long-distance travelers, many of them inured to large locomotives, the S-1 still looked like a great white shark in a school of tuna. The engine, plus its tender, measured 140 feet in length and weighed 1,060,000 pounds. It carried more than 21 tons of coal and 24,230 gallons of water. Its driving wheels were 7 feet in diameter, and the 6,500-horsepower engine was capable of pulling a 14-car, 1,200-ton passenger train at more than 100 miles per hour. The three top locomotive builders in the country—American, Baldwin Locomotive Works in Philadelphia, and Lima Locomotive Works in Ohio—submitted plans for the locomotive. The prototype for the S-1 was built in the Pennsy's test shops in Altoona.

Altoona, Pennsylvania, sprung up as a city because the Pennsylvania Railroad needed it. The town became a center for engine and car repairs because of the city's proximity to a gradual-rise mountain pass called the Horseshoe Curve, which allowed freight and passenger trains to make the steep climb over the Allegheny Mountains. The town grew when Pennsylvania Railroad's Alexander Cassatt, superintendent of machinery and motive power (and the brother of Impressionist painter Mary Cassatt), recommended that car and repair shops be located near the Horseshoe Curve. Altoona also served as the railroad's testing area. The facilities in Altoona tested everything the railway used, from locomotive boilers to whisk brooms.

Raymond Loewy repeats his most iconic pose, perched atop Pennsylvania Railroad's S-1 streamlined locomotive, the largest steam locomotive ever built. The engine, one of a kind, was scrapped in 1949. (Raymond Loewy™/® by CMG Worldwide, Inc./www.RaymondLoewy.com)

The Altoona shops did more than testing, however; prototype loco- motives were designed, built, and maintained there, although mass- production engines were built elsewhere. The locomotive testing plant in the city was exhibited at the 1904 Louisiana Purchase Exhibition, held in St. Louis, Missouri, dismantled after the exhibit closed, and completely rebuilt in Altoona. The Pennsy's experimental engines such as the S-1 were built, tested, and approved in this company town.

"They chose a special crew of men—hand-picked—that they called the 'blue-ribbon gang,' " recalled one retired railroader who worked on the S-1. "They weren't just run of the mill." Few components of the S-1 could be called run-of-the-mill. Nearly every part used in the new loco- motive was tooled and produced in the Altoona shops. The frame—the largest locomotive frame ever cast—measured 77 feet, 9½ inches. The frame was machined to exact specifications, and each hole drilled had to be as "bright as a gun barrel," according to one laborer. A lathe operator remembered, "They wanted everything polished and we spent hours and hours on that. We groused about the polishing because we could make a lot more money doing piece work than shining those parts all day."

The S-1 made its debut at the 1939 World's Fair in New York. The theme of the fair was "Building the World of Tomorrow." Loewy had been hired to oversee one of the three main themes of the fair—transportation. He was consulting designer for the main exhibit in the Chrysler Motors building as well as the entire exhibit in the railroad building. The railroad building was an eight-story dome containing rolling stock, dioramas, and several theatrical productions. Loewy was a consulting designer for parts of the exhibit, but his major contribution to the rail display was the trains and engines he designed. The display housed an animated diorama that detailed railroads, mining, smelting, and timbering oper- ations, complemented by another exhibit that showed how rail systems served the public. Perhaps the most ambitious showcase for the rail- roads came from artist Griffith Bailey Coale, who created a live diorama featuring idealized male actors holding scaled-down versions of engines and railcars.

A Loewy locomotive also appeared in one of the most popular theat- rical productions of the fair, "Railroads on Parade," an open-air extrav- aganza that used actual rolling stock to tell the century-long history of American railroading and transportation. Theatrical producer Ernest Hungerford crowded trains, wagons, boats, 250 actors and dancers, and livestock onto the stage of an outdoor amphitheater, all activity scored to the music of Kurt Weill. The spectacle was staged in sixteen scenes, each depicted by the passing of a train, wagon, boat, or coach. The production included views of the minuscule DeWitt Clinton steam engine to the reenactment of driving the Golden Spike to connect the Central Pacific and Union Pacific railroads. The brilliantly conceived climax coupled extensive choreography featuring dancers clad in avant-garde uniforms of the future with the simultaneous entrance of two streamlined engines. The sleek *20th Century Limited*, designed by Henry Dreyfuss, entered stage right as Loewy's streamlined redesign of the K-4, also known as Engine 3768. The K-4 redesign allowed Hungerford to couple two classic forms into a culminating image of the future of transportation.

Meanwhile, nearby, the future of transportation was displayed on what seemed to be a stand-alone section of tracks. The massive S-1 engine ran every hour on the hour, thanks to an intricate treadmill designed by Keller Barry, a machinist at the Altoona shops. The engine could reach speeds of 60 miles per hour on the treadmill. The engine's labors at the World's Fair were its only appearance in New York in its lifetime. To get to the fair, the engine was towed at low rates of speed into Long Island. Because the S-1's tremendous length made it impossible to negotiate the multicurve tracks leading to the East Coast, Pennsylvania Railroad officials delivered the engine to the fair site by taking it on a tortuous route that crossed several city bridges and required the temporary removal of the electric third rail on Long Island's commuter tracks. After the World's Fair, Loewy, who had been asked to write a book on the aesthetics of locomotives after his K-4 streamline makeover in 1937, was hailed as the greatest train designer of his day.

The S-1's fate after the World's Fair never reached the heights predicted for it by rail executives and the enthusiastic crowds. The terrible beauty of the S-1, designed to lure more rail passengers into Pennsy trains, was an initial success. After the engine left the World's Fair site in 1941, the engine pulled the *Broadway Limited* on its daily runs. "It sounded just like a jet airplane," recalled a retired brakeman who rode on several trains pulled by the S-1. "There was a roar at first, and then it just sounded like a big whooshing sound in the distance. Later on, I went into the cab and it had so many gauges it looked like an airplane."

Unfortunately, the engine never traveled east of Pittsburgh, where the Pennsylvania Railroad's most lucrative routes were. The leviathan

Raymond Loewy assumes a relaxed pose as he leans against a Studebaker car parked next to the S-1 steam locomotive. The giant locomotive engine was displayed for more than a year at the 1939 World's Fair. The Pennsylvania Railroad designed a special treadmill that allowed the engine's gigantic wheels to turn in place so that crowds could see the locomotive in operation. (Raymond Loewy Archive, Courtesy Hagley Museum and Library)

machine was simply too large. The engine was constructed on a rigid frame, which meant it could not make tight turns. The engine was even too large to fit on a turntable, which meant rail yards had to use Y-tracks to turn the locomotive around. The S-1 pulled trains along the relatively flat tracks from Chicago to Crestline, Ohio. In use, the S-1 also had problems with wheel slippage, whereby the power produced by the massive engine created traction problems unless engineers slowed the engine during starts. After going into service, most of the engine's distinctive skirting had been removed to make maintenance easier. The sleek look of the engine had fallen to grime and convenience. By the mid-1940s, it became clear even to stubborn Pennsylvania Railroad executives that diesel engines were the future of railroading. Although the railroad's unswerving faith in steam power would last until the last steam engine was mothballed in 1957, the S-1's era lasted just ten years. In January 1949, it was taken out of service and cut apart for scrap metal.

Loewy would go on to design more locomotives for the Pennsylvania Railroad. His most successful steam design came in 1942 with the T-1 freight engine. Built at the Baldwin Locomotive Works, the engines were originally designated to pull fast passenger trains. Painted dark green and sporting a fine gold line that began low on the running board and continued through the gigantic tender (it held 20,000 gallons of water), the distinctive engine looked much brawnier than Loewy's other Pennsy designs. The jutting prow looked like a balled fist—ready to blast through obstacles. Design historian Paul Jodard called it "Rugged Moderne." Fellow designer Otto Kuhler called it "beautifully streamlined, with an impressive sharknose." The railroad would go on to build fifty of these large duplex engines. Weighing just below 1 million pounds, the T-1 generated 108,625 pounds of tractive effort. This design sculpted the engine's immense boiler into a "shark's nose" prow that was gently rounded to hold the solitary headlamp. As he did with the S-1, Loewy used a jutting platform to visually balance the boiler. Instead of speed lines, however, Loewy used three holes to give the T-1 platform some visual pizzazz. The portholes were probably inspired by automotive decorations used in early cars, including the 1919 Biddle, the 1919 Greyhound, and the 1926 Shaw. In 1948, Buick reintroduced the portholes (perhaps inspired by Loewy's T-1) and used that design trope to "brand" its models up until the 1990s.

The T-1 locomotives were hailed by the railroad as the solution to encroaching diesel engines. Ralph Johnson, Baldwin's chief engineer, described the T-1 at the New York Railway Club: "These locomotives will outperform at 5,400 horsepower at all speeds above 26 miles per hour and if given comparable facilities for servicing and maintenance, will do the task more cheaply." In keeping with Johnson's sunny assessment, the locomotives were incredibly fast, but they also were incredibly expensive to operate. Maintenance costs were high, and the driving wheels had a hard time gaining purchase on rails while starting to pull a fully loaded train. What the T-1 gained in speed, it lost in the shops, a problem that worsened as the engines aged.

The shark-nose style of the T-1 was used for Loewy's later designs for Pennsylvania Railroad diesel locomotives, and some claim it inspired the design for the World War II P-40 fighter. Other railroads also adapted the shark's nose for their own locomotives. Loewy's firm continued to design for the railroad as well. He would oversee designs for dinner

A precisionist-style painting of the T-1 freight locomotive, Loewy's last streamlined locomotive. Its distinctive "shark's nose" is echoed in the look of the diesel freight locomotives used today. The Pennsylvania Railroad built fifty of these steam locomotives at the end of the steam era. (Raymond Loewy™/® by CMG Worldwide, Inc./www.RaymondLoewy.com)

menus, toothpicks, matchbooks, signal towers, coffee cups, an entire bridge over the Potomac River, and a ferryboat for a Pennsy subsidiary.

The last gasp for streamlined high-speed engines came from—surprisingly—General Motors and Loewy's styling rival Harley Earl. The Aerotrain emerged from Earl's Special Projects Studio in 1955, created in response to requests from several railroads—including the Pennsylvania, New York Central, and Union Pacific—to develop a lightweight, high-speed passenger train. The designers came up with two prototypes; each had ten passenger cars pulled by a single engine. Despite the train's relatively light weight, the futuristic locomotive was not enough engine to pull it. Two downsized forty-passenger cars weighed less than one standard eighty-person railcar. GM also cut out headroom and used aluminum instead of steel in many applications. Designed in large part by Chuck Jordan, whose automotive gems include the 1959 Chevy Impala, the 1958 Corvette, and the 1959 Cadillac Eldorado Biarritz, the Aerotrain had a sleek otherworldliness that recalls fantasy trains and the excessive chrome and aircraft-inspired decoration that overloaded many

1950s automobiles. The locomotive features a host of GM car styling cues, many inspired by Earl's concept car, the 1951 Buick LeSabre. Like the LeSabre, the engine featured a wide oval "intake" mouth mimicking the manta ray intakes of fighter jets. Within the oval were multiple headlights looming out over the cowcatcher. The engine also featured a wraparound windshield. The final stroke of genius—or, more accurately, overreaching—was the trailing observation car, which sported taillights and abstracted fins that suggested to many observers the rear end of a 1950s Chevy Nomad station wagon. The Aerotrain was a procession of GM bus bodies modified as railcars pulled by a 1,200-horsepower diesel engine. The Pennsylvania Railroad leased the train in 1956, as did the New York Central, but the lightweight cars, using suspensions designed for highway buses, did not run smoothly at high speeds. The smaller cars, much less roomy than standard train cars, made many passengers feel claustrophobic. Other railroads tried out the Aerotrain, but the result was the same: bumpy, jarring rides. By 1957, the trains were sold to the Rock Island Railroad to be used for low-speed commuter service. After just ten years, the Aerotrains were mothballed, the same fate that befell Loewy's S-1.

At the end of the 1960s, the New York Central experimented with a jet locomotive. Engineer Don Wetzel, of the Cleveland Technical Center, adapted two General Electric jet engines and mounted them above the engineer's compartment at the front of the train. The jerry-rigged locomotive set a land-speed record of more than 183 miles per hour. The jet locomotive was retired, but Wetzel used it later to test a rail snowblower. The twin jet engines blasted away all kinds of snow, in addition to the railroad ties and gravel beneath them. Eventually, the New York Central engineers adapted the jets into a prototype snowblower that is still used on rail engines across the country.

Loewy's locomotive designs are his most celebrated works for the Pennsylvania Railroad, but most of Loewy and Associates' work had nothing to do with rolling stock. Billing records show that the company worked on corporate logos, advertising, and plaques; created exhibition signage and the exhibits; chose paint, textiles, and carpeting; and even recommended menu items for dining cars. The company also delivered many designs for interiors of railcars, railway stations, and other Pennsy real estate. The clash of the corporate cultures is capsulized in a reminiscence by designer John Ebstein, who had been assigned to photograph the Pittsburgh railroad station. Ebstein couldn't make decent exposures because of smog obscuring the station. "I went back to New York totally in disgust." Ebstein was even harder on Altoona, remembering it as "the dirtiest place. We hated to go there. The smoke and the coal dust. The hotel rooms were dirty. I do not recall whether Loewy spent any time in Altoona."

One of his most complete designs was for the interior of the passenger train the *Broadway Limited*, which ran every day from Chicago to New York City. It was the flagship train for the railway. The look of the train was designed to be unforgettable. Loewy collaborated on the interior with Paul Cret, head of the University of Pennsylvania School of Architecture. The interiors of the train were meant to mimic the sumptuous interiors of exclusive clubs or hotels.

Cret's work on Art Deco interiors and architecture was subdued yet effortlessly memorable. His stark Art Deco architecture includes the Folger Shakespeare Library and the Federal Reserve Building, both in Washington, DC. He also excelled at train-related design, creating the look of Cincinnati's Union Terminal in 1933. Cret, who was born in Lyons, France, graduated from the École des Beaux Arts in 1903 and emigrated to the United States after serving in World War I. Unlike Loewy, who came to America ready to join the business world, Cret became a professor of design at the University of Pennsylvania, a post he held until 1937. A shared history with Loewy made for a productive partnership. The *Broadway Limited* was one of the first trains in the nation to eliminate upper and lower berths in passenger cabins, preferring a design using just lower berths. He also tackled a variety of train interior assignments, including the *Pioneer Zephyr*, the Santa Fe's *Super Chief*, and the New York Central's *Empire Express*. Cret used design motifs inspired by the Pueblo Indians for the interiors of the *Super Chief*, as its route took the train through tribal lands. For the *Broadway Limited*, Loewy's firm designed the interiors of the passenger cars. Cret and Loewy each designed a dining car for the main train and other routes.

By that point, passenger train travel had ebbed, and steam-powered travel was nearly nonexistent. Most of the locomotive builders after Loewy and Dreyfuss followed the styling motifs of General Motors engines, which essentially used the boxy shape and a cab with a snubbed, rounded nose. In 1948, the Baldwin Locomotive Works hired Loewy to design new diesel prototypes, which featured an angular engine with an extremely angled shark-nosed prow. That basic look is still in use today and provided a sharp contrast to the smoothness of the GM models.

PERHAPS NO STRONGER ENDORSEMENT of Loewy's mastery of locomotive design came in the form of a writing assignment. His 1937 book *The New Vision: The Locomotive* was the second in a series of "style manuals" produced by Studio Publications in 1936 and 1937. The first author to write one of the manuals was architect Le Corbusier, who wrote *Aircraft*. Le Corbusier, who celebrated the machine over human factors, was unconcerned about style trends. Sir Reginald Blomfield famously said that the architect's interiors seemed to be designed for "vegetarian bacteriologists." Other books in the series were planned, but World War II put an end to the enterprise. Loewy's book is a collection of photographs of streamlined engines accompanied by the designer's brief but cogent notes. He could not conceal his disdain for diesel locomotives and saved most of his praise for the steam trains of his European youth and those he or his competitors worked on. Still, the author could not resist adding a few lines of entertaining patter. "My youth was charmed by the glamour of the locomotive. Unable to control an irresistible craving to sketch and dream locomotives at the oddest moments, it was a constant source of trouble during my college days. Later, as a young man, it led to my complete oblivion as a dancing partner."

Dancing talent aside, the longest-lasting legacy for Loewy continues to be the massive engines that he designed. His relationship with the Pennsylvania Railroad lasted decades and produced countless designs. When Loewy's papers were put up for sale after his death, a group of American designers banded together to buy them for the Library of Congress. In that collection is a drawing Loewy made as a teenager in France in 1911. The drawing, of a train of course, holds the same vibrant signature that graced every design to come out of his offices—whether Loewy personally designed the product or not.

6

Constructing an Image
while Building a Business

At two critical points in his career, in 1933 and again in the mid-1940s, Loewy decided to gamble with his new design business. The first occurred during the economic downturn after the stock market crash of 1929. He was bereft of most of his clients and, according to his autobiography, about $125,000 in the hole. "So I rented a swank office [in 1933] on the fifty-fourth floor of 500 Fifth Avenue, furnished it sumptuously—and made a sparkling success of it until 1935." Business improved rapidly and necessitated a move to a larger Fifth Avenue address—this time a penthouse suite and once again in the "right" neighborhood—where he served about a dozen clients. Such strategy reflected savvy anticipation—that companies would test marketing waters by hiring consultants rather than expanding payrolls or putting forth the public appearance of success. As public relations pioneer Benjamin Sonnenberg Sr. would say, "Always live better than your clients." Loewy had his own take. "Regardless of what people may tell us, or experts may write, I stick to my basic philosophy. It is a simple one: I believe that in our country there is always a chance of success for everyone (a) Who knows how to do a thing well (b) Delivers it on time (c) Sticks to his word."

Loewy seemingly built his business with nary a misstep. Yet few business records from the company prior to the 1950s are in any of Loewy's archives, so it is impossible to know whether his path was as smooth as he claims. As always, Loewy is the one establishing the story for the reader, and he presents the rosiest scenario. Almost from the beginning,

Loewy used a corporate model to form his organization. Beginning in the 1930s, he separated the firm into departments. The size of the company allowed Loewy to take on more work and, perhaps more importantly, let him emphasize the sheer size of the organization as a selling point to the major industries and conglomerates he sought as clients. Similarly, he was able to recruit designers by offering them a wide diversity of design assignments and an opportunity to "cross-train" into new specialties.

Initially, Loewy's firm had three divisions: Products, Transportation, and Packaging. In 1934, he hired a young architectural engineering graduate whose first job was in the General Motors Styling Division. A. Baker Barnhart, known as "Barney" to everyone at the firm, worked on all the company's early jobs, including the GG-1 locomotive, the Sears Coldspot refrigerator, and the Hupmobile. Loewy referred to Barnhart as "the real American thing." In essence, Barnhart was the linchpin for customer service within the company. He was the design liaison with every important Loewy client: Studebaker, Pennsylvania Railroad, Greyhound, Nabisco, Frigidaire, Coca-Cola, and Trans World Airlines. Barnhart was the hearty Main Street counterpoint to Loewy's European savior faire. Loewy wrote of him: "A fellow of taste, he is not only talented but everybody likes him. Whether men on our own staff, clients, or suppliers, everybody agrees Barney is a grand guy. Sensitive and keen, he is at his best in complicated situations where the human angle is foremost as a factor." Translation: He's the guy who talks to the clients.

"He was closest to the principal clients, many of whom were his personal friends," wrote Betty Reese, the public relations executive for Loewy's company. "The Loewy people, too, found him sympathetic and knew that his intercession with Loewy on their behalf would be confidential and effective." In effect, Barnhart was the company's sociable center in comparison to Loewy's somewhat distant and reticent patriarch. Barnhart hired a second designer early on, when the company began winning a series of product design contracts. Carl Otto, who had trained in the styling studio of General Motors, was a product designer who also seamlessly integrated his skills with both Loewy and staff. Otto and Barnhart hired more designers and staff, managed most of the projects, and ensured Loewy's design vision.

Perhaps the most important hire came on in 1936, when Loewy accepted an assignment to redesign the stationery department of a New York City department store and decided to take on retail space design, a new niche. William Snaith, trained as a stage designer, had studied architecture in Paris and New York. Snaith would spin off his initial assignment into a co-equal full partnership and the establishment of Loewy's Store Planning and Design Division. Snaith headed the largest design operation in the firm, which, in the years after World War II, eventually overshadowed the company's product and industrial design divisions. Ultimately, Snaith had his name placed in equal billing with Loewy as a co-owner of Loewy and Associates. The firm grew to five partners, including John Breen, a business manager recruited from the advertising department of the *New York Times*, and Jean Thomson Loewy Bienfait, Loewy's first wife, who would stay with the company through their divorce and her remarriage.

Loewy met Thomson in 1929 and was taken with her, a woman who had been educated in France and spoke French fluently. They were married in 1931. She was an integral part of the firm, helping Loewy cultivate business contacts and creating a social atmosphere for business. Effortlessly stylish, she worked with her husband as hostess, companion, and sounding board. She played much the same role that Viola Erickson Loewy, the designer's second wife, would take on after the war.

The firm instituted a profit-sharing program as well, which may have helped assuage the boss's penchant for grabbing credit. On the other hand, having Loewy as the public face of the company meant that the firm's other employees rarely had to make cold calls. Once the company was established, most commissions came in "over the transom."

Reese describes his management style: "During the working day he toured the offices, making suggestions to division chiefs, leaning over drafting boards to compliment certain drawings or sketches, and then retiring to his private office to plan strategies for selling, selling, selling!" Typically, Loewy made initial presentations to new clients, then, once the account was in hand, he would turn over daily client maintenance to one of his division heads. If there were to be major presentations for products, such as a new car model or product line, Loewy would almost always travel to the client's headquarters to make that presentation. Early on, Barnhart, Snaith, and Otto accompanied Loewy on his "lonely trips to Midwest manufacturers." As the clients became comfortable with whichever partner was best qualified to service the account, Loewy would reduce his client visits unless a problem arose or a client requested his presence. His last words on the subject were succinct:

The partners at Loewy and Associates circa the 1930s. *From left*: Raymond Loewy; Bill Snaith, who developed the company's retail design operation; Jean Bienfait, Loewy's first wife and public relations representative; A. Baker "Barney" Barnhart; and John Breen, business manager. (Library of Congress, Prints and Photographs Division, Visual Materials from the Raymond Loewy Papers [Reproduction number cph.3c04137])

"Nobody can sell industrial design but an industrial designer. We tried salesmen. We tried them in New York. We had all kinds, very high-priced fellows they were. Absolute flop."

Loewy's risky expansion as the country emerged from bad economic times was triggered by the 1934 Sears Coldspot account, which generated several model updates over the life of the contract. By 1938, Loewy had hired eighteen designers and expanded the business well beyond the Sears account. By 1941, he oversaw fifty-six design professionals. At the height of the firm's success, from roughly 1947 to 1965, the company employed 150 to 180 people in permanent positions. Depending upon the number of contracts, the company also employed nearly two dozen freelance draftsmen, model makers, and detailers.

The company did not ascribe to a particular design aesthetic, such as the ergonomic work of Henry Dreyfuss or the streamline theories of

Bel Geddes. Loewy incorporated styles he felt fit the sales potential of the product, resulting in designs that echoed Art Deco, Art Nouveau, Bauhaus, and Streamline Moderne. His design philosophy centered on simplification and giving the client a product that stood out in the marketplace, yet it appealed to a large demographic of consumers. His criteria in hiring was similar, focusing on a designer's ability to draw a wide range of subjects and proficiency in a variety of drawing techniques. He also needed speed. Recruits were tested on the boards for several weeks, and if they could not produce, they did not stay. "We can't spend all day on a single drawing, unless it's a finished rendering or a presentation drawing," quoted Reese.

Reese described her boss's presentation protocol: "Loewy prepared for his presentations like a boxer preparing for a big match. He wouldn't eat; he'd be at the office hours before everybody else; he'd practice every word, every joke. When the rest of us straggled in bleary-eyed, he was there, fit for a championship fight. Then, if he was nervous about his speech, he'd start off by saying charmingly, 'Please forgive my terrible French accent, I've only been in the country thirty years.' "

All design drawings were executed by staff artists. What Loewy brought to the company was an exacting eye for unique and salable designs and an extrasensory talent for knowing the consumer marketplace. Snaith characterized Loewy's sense of the consumer landscape as "an unerring vulgar taste." Snaith was not denigrating his partner; he meant that Loewy had a feel for what would appeal to prodigious masses of customers. Design historians Stephen Bayley and Terence Conran wrote that "Loewy combined in one stage-managed personality the flair of Bel Geddes and the practicality of Dreyfuss."

All finished design drawings were signed, typically with "Loewy," "Raymond Loewy," or "RL," depending upon the account. Loewy's business signature did not resemble his actual signature. Like Walt Disney, who had a generic signature designed by the cartoonists who worked for him, Loewy's graphically bold signature was a product of the company. All drawings, save for the signed final design used in the evolution of each assignment, were destroyed. Because of this practice, historians have subsequently had trouble tracing the evolution of Loewy designs from conception to prototype. Jay Doblin, a designer hired in 1939 as an office boy and who later supervised the entire office, once said, "The best designs always ended up in the wastebasket."

Over time, through his own personal research and development, the designer formed a public persona that was equal parts entertainer and impresario. "Loewy was a great manipulator—one of the best—and he learned how to manipulate the public's image of himself," was Betty Reese's clear-eyed assessment of her boss. "He decided at about the age of twelve or thirteen to create this character called Raymond Loewy and that's what he perfected. As much as he adored cars, I only saw him actually drive once or twice; he always had a chauffeur. He would meet these Midwestern company presidents at the airport in his extravagantly beautiful suits [with] his chauffeur and they would be immediately floored by his style."

The second critical point, an expansion in the mid-1940s for Loewy's company, perfectly positioned his designers for success. The postwar consumer market was the largest in history. In 1946, personal consumption was 20 percent higher than in 1945 and 70 percent higher than in 1941. The exploding birth rates that followed the war created an instant market for all kinds of consumer products. In 1960, there were 55.7 million children under the age of 15, the largest single generation in US history. Unlike other postwar economies, there was no consumer slowdown after World War II. Consumers searched for more expensive and diverse products. In 1944, "as an expression of confidence and gratitude to my key men, I took them as partners and we became Raymond Loewy Associates. It gave further élan to the organization." It also gave further peace of mind to Loewy, who did not have to worry about numerous defections.

Jay Doblin—who worked as executive designer for such accounts as the Pennsylvania Railroad, Greyhound, and Frigidaire and went on to become the director of the Illinois Institute of Technology's Institute of Design from 1949 to 1959 (he also established Unimark, a corporate identity consulting firm)—said Loewy was willing to take a chance on a designer to see if the new hire was up to the challenge. "I started working for Loewy as an office boy in 1939 when I was still at the Pratt Institute," Doblin recalled. "I had two introductions, one to Loewy and one to Norman Bel Geddes. I first went to Bel Geddes' office and sat in the lobby and waited for ages for him to show up. Finally I said to hell with it and went across the street to Loewy's [shop] and got a job on the spot."

By 1939, PUBLICITY FROM THE World's Fair catapulted Loewy and his peers to the forefront of popular culture. In addition to Loewy's dazzling

S-1 engine and the theatricality of the Railroad Building, Loewy collaborated with architect James Gamble Rogers on the Focal Exhibit in the fair's Transportation Zone. Sponsored by the Chrysler Corporation, the building was geometrically designed by connecting a long, rectangular central hall with an oval central axis. Each side of the central oval featured towering pylons designed in the Streamline Moderne style.

The building's main hall was an air-conditioned oasis seating 360 onlookers. The crowds used the theater to watch a three-dimensional movie that detailed the step-by-step construction of a Plymouth automobile. Upon exiting the temperate theater, visitors stepped into the "Frozen Forest," which featured Plymouth, Chrysler, De Soto, and Dodge cars parked amid a jungle of artificial palm trees. Banks of Chrysler Airtemp air conditioners lowered the temperature in the room to wintry temperatures, in some cases forming frost on the ersatz fronds. The sparse but effective demonstration of man's ability to alter a humid summer into a winter wonderland was inspired by Loewy's earlier work on window displays.

The Chrysler Pavilion at the 1939 World's Fair, designed by Loewy and Associates. The more celebrated pavilion was the General Motors exhibition, which included Norman Bel Geddes's *Futurama* display, but Loewy's work for the smaller Chrysler display was innovative as well. (Raymond Loewy Archive, Courtesy Hagley Museum and Library)

Loewy's participation in the 1939 World's Fair generated so much press that Loewy realized he needed to bring in an expert to market the firm. He hired Betty Reese in 1940. Her insight was simple: put Loewy at the front of the pack and create events that allowed him to sell the services of the company. "I never got credit for anything I did at Raymond Loewy, nor did anybody else. But I didn't care. The man was the greatest salesman to ever hit the design business," Doblin wrote in the 1986 Loewy memorial issue of *Industrial Design*. "He made it possible for us to work on marvelous accounts, learning, having fun and getting paid for it. That's why none of us resented him grabbing one hundred percent of the credit."

The written record on Raymond Loewy's various public and private personas are best captured by his loyal publicist Reese. Betty Reese grew up in the heart of Ohio's corporate center and manufacturing belt in Cleveland, which she left shortly after her high school graduation to work as a vaudeville dancer. She used her earnings to pay her way through Case Western University, earning a bachelor's degree in English. After graduation, she moved to New York City and found work as a radio actress. She also worked the 1939 World's Fair as a puppeteer, where she no doubt encountered some of Loewy's designs. She segued out of entertainment and into public relations by getting a job in the publicity department of the fair, moving on later to the public relations arm of the BBDO advertising agency in Manhattan.

Even the public quotes from Viola Loewy were not as revealing as the remembrances Reese wrote for the Loewy memorial issue of *ID Magazine* in 1986 and in an essay she wrote for *Raymond Loewy: Pioneer of American Industrial Design*. "My initial impression of the man was that he was terribly refined and elegant. He told me right away he wanted the cover of *Time* magazine. I was a little taken aback by that. It took me eight years, but I did get it for him!" she recalled. "I was a newspaperman's daughter, so I grew up learning a lot of the tricks of the trade, and I must say Loewy picked them up quickly. For example, I told him that whenever photos were taken of him with some bigwig to always squeeze in to the right of the guy, because in captions, they identify people from left to right [still common journalism practice], and Loewy's name would then always come first. He learned all the tricks and worked hard at it."

Reese never missed an opportunity to put Loewy and a product in front of a camera or reporter. She counseled her client in the art of the staged event. Eventually, the New York art community took notice of Raymond Loewy. In 1937, Loewy collaborated with architect Lee Simonson to create a "designer's studio" for the Metropolitan Museum of Art. The minimalist aesthetic of the space, with walls lined by horizontal moldings, horizontal windows finished in a semicircle, and metal-framed furniture, is a glimpse into the Streamline Moderne look of the 1940s. Loewy described his idea for the room as "a clinic—a place where products are examined, studied and diagnosed." The antiseptic look of the room is achieved by using Formica walls, blue linoleum floors, and indirect lighting. The office is further decorated by a model of Loewy's Hupmobile and framed sketches of the *Princess Anne*, a ferry Loewy designed for a subsidiary of the Pennsylvania Railroad. The visual joke

Raymond Loewy poses in the Metropolitan Museum of Art's 1937 display of the "Designer's Studio," a collaboration with architect Lee Simonson. The spare aesthetic of the space, with walls lined by horizontal moldings, horizontal windows finished in a semicircle, and metal-framed furniture, is a vision of "a clinic—a place where products are examined, studied, and diagnosed." The office is not functional. It is tiny, cramped, and missing the storage space any businessperson would need. (Library of Congress, Prints and Photographs Division, Visual Materials from the Raymond Loewy Papers [Reproduction number cph.3c04127])

is that the office is hardly functional. It is tiny, cramped, and bereft of the storage space any businessperson would need. The office is, in the end, a stage set, created by a man who had no connection to the theater yet constantly played the role of the cultured, cosmopolitan aesthete.

The preserved image was one of Loewy's most important photo opportunities. Loewy poses leaning against a table in the middle of the room. He is dressed in slacks and a jacket, hands clasping a casually crossed leg. His pose is of the country squire perhaps influenced by the movie-star stills made famous by George Hurrell and Cecil Beaton.

In 1940, buoyed by the reception given for his work at the World's Fair, Loewy was invited to return to the Metropolitan Museum of Art to construct a designed room for the *Contemporary American Industrial Art* exhibit. The "Room for a 5-Year-Old Child" was conceived and executed solely by Loewy and the firm, including all textiles and wallpaper. (At the time, Loewy had no children.) The press release announcing the design stated that typical children's rooms were "grossly over-decorated, cluttered with fabrics, wallpapers and gadgets in the ugly 'Alice in Wonderland' tradition." Two walls of the Loewy room were paneled in pine, and another was covered with a pink textured wall covering. The floor was rubber tile, which was decorated with stars. The furniture featured such materials as Formica, Nucite, aluminum, and white plastic. The bedspread for the child's bed was embroidered with a horse, house, and rooster, to acknowledge that an actual child would sleep there. Loewy's most predictable addition to the room was a toy automobile, featuring a streamlined body and a white interior. Its exterior colors were white, gray, and silver, perhaps using leftover paint from a Greyhound project. He also designed a small stove with brass doors. Loewy's press release stated, "The designer believes that the child should get acquainted early in life with the beauty and dangers of fire." Indeed.

The events Betty Reese arranged could be official, such as the unveiling of a new locomotive design, or impromptu, such as the designer having his photo taken as he disembarked an airliner with a fellow passenger, a reigning beauty queen. Reese accomplished Loewy's top goal, to get him on the cover of *Time*, on October 31, 1949. Although the postwar feature story profiled a man at the height of his influence on his chosen profession, well after his struggle to build and expand the company, the story essentially reinforced and publicized the Loewy persona as an urbane sophisticate who created a design empire through the force of his own good taste. That story, and another published the same year in *Life* magazine, mirrors the same "creation stories" that Loewy later put to paper in *Never Leave Well Enough Alone.*

The cover revealed a line drawing by Loewy as examples of his products and design orbited his head in what can only be described as a postmodern halo composed of the S-1 locomotive, the Studebaker Commander, the Lucky Strike package, a Frigidaire stove, a Hallicrafters radio, and an egg (Loewy was quoted in the story as saying the egg is the perfectly designed package). The tone of the story is a publicist's dream, as the writer sets the scene by describing Loewy waking up, switching off the Loewy-designed alarm clock, shaving with a Loewy-designed razor, and brushing his teeth using a Pepsodent tube "modeled by Loewy." The story's headline, a concise endorsement of design as sales tool, could have been written by either Reese or Loewy: "He streamlines the sales curve."

Designer John Ebstein, who worked airbrush magic for Loewy's transportation division and helped Studebaker conceive the Avanti sports car, said, "He was, when he wanted to be, irresistibly charming. He also had an intuitive gift for instantly seeing what was good or bad in a design. He knew what would sell. He couldn't draw a straight line, but he had an uncanny feeling for what a product should look like."

BY THE MID-1940s, Loewy's firm had multiple accounts, and designers were free to move between accounts or to take assignments in other offices if the product line they normally worked on was experiencing a business lull. The size of the firm also allowed designers to concentrate solely on design work. Often at smaller agencies, designers had to take on additional duties ranging from office work to public relations. Loewy paid his designers well, and this generosity, coupled with the ability to concentrate on individual specialties, cultivated loyalty among employees. Many of the designers hired by Loewy stayed for decades, and those who left often opened firms specializing in a narrower range of design, using expertise honed working on Loewy accounts. Designers experiencing burnout also could be temporarily shifted onto other accounts. The Nabisco account, known within the organization for corporate micromanaging, regularly rotated designers. According to company records, more than five hundred designers worked on Nabisco during its duration as a client.

"Many designers were happy for the experience," wrote Ebstein about working for Loewy. "There was enormous enthusiasm in the office, where we were all learning so much. We would work incredibly long and hard hours—night and day sometimes—but with such excitement. We grew up with the field of industrial design. I remember [hiring] our first engineers, our first marketing division, our first packaging group. It was all terribly exciting."

The overarching mission of the Loewy shop was to provide designs that could be modified and improved over the life of the product. Called "planned obsolescence" by designer Brooks Stevens, one of Loewy's later rivals, the practice of creating products that reached an optimal design by incremental changes was loved by consumers and manufacturers and loathed by art critics and design purists. The central idea of pure design, best characterized by the International Style in architecture and by elements of the Bauhaus aesthetic, was to achieve perfection in a product's essential form and leave it unchanged. Stevens, who was sensitive to attacks on the phrase he coined, said in 1958, "Our whole economy is based on planned obsolescence and everybody who can read without moving his lips should know it by now. We make good products, we induce people to buy them, and then the next year we introduce something that will make those products old-fashioned, out of date, obsolete. We do that for the soundest reason: to make money."

To consultants serving the American manufacturing community, helping the company make money was the priority. Hewing fanatically to an unbending theory of design was a recipe for a disgruntled client roster. Reese details Loewy's innate sense of keeping clients happy: "Loewy made sure to orchestrate the clients' first view of his designs. If it was a car, for example, he would show it to them first from an impressive distance, from the end of a very long studio. He had learned that if the clients were allowed to get close to the car in a casual way, they would start picking it apart. He wouldn't let them get near it until they had experienced it as a complete and finished design."

The same sense of distance held true in Loewy's business life. A man of many acquaintances but relatively few close friends, Loewy kept his employees and most acquaintances at arm's length. One person Loewy allowed to get near to him was Viola Erickson, a public relations executive, whom he courted and married in 1948. Viola was a striking and outgoing woman who perfectly complemented her husband. Like Jean

Thompson Bienfait before her, Viola Loewy was expected to accompany her husband on social and business occasions and to interact seamlessly with the parade of manufacturers, businessmen, and Rotarians capable of offering the Loewy company commissions. He recognized her value as a business asset. Although she was never an official partner on the firm, she carved out a significant role, handling clients, negotiations, and contracts. In his memoir, Loewy tells of his obsessive "training" before major presentations. His audience and key adviser in these sessions: Viola. "When I first met Raymond, I found him much too flashy. He always wanted to make an impression on people and I must say he succeeded: he was creative, brimming with ideas—and he knew how to sell himself," she recalled.

By creating himself as a "brand," Raymond Loewy allowed the office to function creatively within the structure of the firm's divisions. As Loewy became the public face of industrial design, a movement that reached its height after World War II, many of his colleagues remained relatively anonymous, even Loewy's original partners. He paid as well or better than any other designer, and the size of his company and client list guaranteed a relatively steady income. If the firm lost an account, Loewy often could move designers to other accounts or find jobs for those who were laid off. Still, only one of the original partners, Carl Otto, decided to strike out on his own, starting his own successful product design firm in 1948.

Some of the designers who began and often ended their careers under the Loewy signature include interior designer Harry Neafie; product designers Fred Burke, Jay Doblin, Clare Hodgman, Joseph Parriott, and Peter Thompson; package and graphics designers John Blyth, Penny Johnson, Gary Kollberg, Ronald Peterson, and Walter Young; and store designers Dana Cole, Andrew Geller, Maury Kley, and William Raiser. Most of the car designers who worked for Loewy also had high praise for the experience.

Loewy recounts an illustrative story in his autobiography that underlined how design work affected more than just consumers and manufacturers. As he made a client visit to Frigidaire Corporation in Dayton, Ohio, the company's general manager, Elmer G. Biechler, invited the designer to dinner at his home, telling Loewy he "wanted to show him something." After spending nearly an entire page detailing the dinner and the wine the two consumed, Loewy describes a car trip to the Frigidaire plant during a shift change. In perhaps the best writing of his entire book, he details:

> No sounding of horns, no brake screeches, only a mighty purr, a feeling—of order, precision, power. As we reached the crest of a hill we could see the stream of red taillights and the stream of white head-lights fading way in the distance. The sprawling plant was ablaze with blue mercury light. Over certain areas the sky was shivering with the blue-white flashes of automatic welding. White, green, red and blue signal lights would punctuate the night. The whole sky was aglow. . . . I was utterly moved by the magnificence of it all. It was like seeing the actual flow of the rich, red blood of young, vibrant America . . . We paused in a quiet spot, and Biechler took my arm and said: "Loewy, my friend, I wanted you to look at all this. You see, when you and your boys work on our problem, in your penthouse office on Fifth Avenue, you may not realize the real importance of the pretty lines you put on paper. You see, every one of these men around us supports a family of four . . . They all live well because they have a job. They have a job be-cause, among other things, your design clicked. In this plant alone—and we have dozens of others all over the world—eighteen thousand men are employed. Eighty thousand dependents! And remember for each man employed at the plant, there are three in the field: sales-men, advertising men, maintenance men, traffic and transportation fellows, warehousers and accountants, dispatchers and repair crews, electricians, statisticians, engineers, draftsmen, etc. That's another sixty thousand. If you add to that another 250,000 for dependents, you get as true picture. More than 320,000 people whose life is directly affected by the success or failure of what you put on paper."

Loewy made that story a cornerstone of his business philosophy, incorporating it into training for new employees. "We never lose contact with reality, and we do not underestimate our social responsibilities. As we have over one hundred active clients on our list, it may well be that the soundness of our designs affects the lives of millions."

As he expanded his client list after the war, Loewy's work for Frigid-aire led to forays into other sections of the kitchen. He signed with Ecko Products, the nation's largest maker of flatware, utensils, pots, and pans. The company's president, Arthur Keating, gave the Loewy firm one mission: keep the product line changing and selling. Eliot Noyes, designer and former curator of the Museum of Modern Art, singled out two Ecko designs for a 1949 *Consumer Reports* article titled "Good Design in Everyday Objects: A Collection of Kitchen Tools for the Flint Brand and an Ecko Best Eggbeater." The yearly Ecko retainer for the Loewy company would eventually total $75,000 (about $859,000 today). Loewy's signature contribution to consumer styles for Ecko came in its home flatware. In the postwar market, Ecko bought out several European manufacturers that had improved the look and reliability of stainless steel. These breakthroughs made it possible for Loewy to introduce contemporary style to flatware that had previously been heavily ornamented. The first entry in the new style line was called Loewy Modern, which Loewy's press release hailed as groundbreaking. Unfortunately, it did not sell particularly well.

Loewy's initial expansion in the late 1930s also meant that the company's headquarters in New York City could not effectively serve its clients. The firm created two major branch offices—one in South Bend, Indiana, to service the Studebaker account, and the other in Chicago, Illinois, to serve the company's accounts with Armour and International Harvester. Under the direction of designer Franz Wagner, the Chicago office expanded to include accounts with the Greyhound Bus Company, United Air Lines, Rosenthal Block China, and Hallicrafters.

The somewhat obscure Hallicrafters design may have been one of Loewy's most influential images. The small company manufactured radios for "ham" radio operators and some consumers, and later made small television sets. For Hallicrafters, Loewy and his designers, led by Richard Latham from the Chicago office, played up the technical interest of the ham operators and reduced the company's offering of commercial radio products. The design, which emerged in 1947, was minimalist, using matte-black casings featuring integrated speakers and large, easy-to-adjust knobs. The controls were organized into groups, and knob settings were color coded. As transistors came into use in electronic appliances, Loewy's modernist take on ham radio sets became the new look for radios, audio components, and to a lesser extent televisions. Later, Loewy created a one-off radio design that further extended many of the details from the Hallicrafters commissions. The bottom half of the radio was executed in black, offset by a clear plastic top half, through which the tubes and wires within the set were visible. The dials were

large, and several of the tuning dials were replaced by aircraft-style tog-
gles. Finally, the directional antenna was encased in a plastic disc. Still
compelling decades later, it never went into production.

The Hallicrafters contract was relatively short-lived for Loewy. Most
designers found that the average life of a design contract for any manufacturer was about sixteen years. As the business grew or ebbed in
accordance with business cycles, the New York office was eventually
no longer the sole focus for Loewy and his partners; the Chicago office
would become the largest profit center in Loewy's organization. "Franz
[Wagner] has a flair for organization and the Chicago branch is a model
of efficient management. Its business curve is a gentle, lovely upward
sweep, a pleasure to the eye," Loewy wrote, recycling a favorite line.

The company also branched out to an office in London in 1934.
The first client for the London office was Sigmund Gestetner, the pioneer client whose initial duplication machine commission launched
the Loewy business. Loewy sent Carl Otto to establish the office, which
was located on the fifth floor of Aldwych House, also headquarters for
Gestetner. Shortly after opening, Otto was awash in commissions and
had hired local designers, including John Bresford Evans and Douglas
Scott, as well as Clare Hodgman, who was sent over from the United
States to work on car designs for Hillman and Roote's Sunbeam Alpine
sports car (most memorably featured in the television series *The Saint*).
Inspired perhaps by his peripatetic boss, Otto was rarely in the office
during the 1930s, preferring to travel to gather new clients. Scott made
sure the office functioned smoothly. The London office grew so rapidly in the five years of its operation that its client list rivaled the North
American operation. Client companies included Lyons, GEC, Electrolux,
Philco, Raleigh, and Gestetner.

The designs coming out of London were not as cutting-edge as those
emerging from the United States, but the consumer demand for ever-
improving products was not as intense in the cash-poor European post–
Great War economy. Loewy's office worked on automotive, kitchenware,
and food packaging accounts.

Some of the more memorable designs to emerge from the office: the
1938 "Housewife's Darling" washing machine, streamlined toasters, and
coffeepots for GEC; the Aga stove; and a minimalist "cooker" created
for Allied Ironfounders that remained unchanged forty years after its
debut. Bill Snaith's retail division also was called upon to create retail

designs for Lyons Tea House. At the height of its momentum, the office was forced to lock its doors in 1939, on the eve of the coming Second World War.

The largest "boutique" operation within the Loewy New York office was the retail store design division operated by Bill Snaith. "Some people who know him well suspect that he is not unaware of his brilliance," said Loewy of his most talented partner. "He has absolutely no design inhibitions. A brilliant partner, Bill is fundamentally an artist, and he has the characteristics of the artist. He is either in the dumps (I mean the real low-down, abysmal dumps) or soaring into stellar space like a hysterical rocket. This characteristic has been known, at times, to be a bit on the trying side, but one gets used to these cycles, the recurrence of which can be pretty well charted." Although few people in the 1950s were knowledgeable about bipolar disorder (then called "manic depression"), it seems that Loewy is describing Snaith as suffering from a type of mania, which the author thought not to be overly debilitating—to Snaith or the firm. Snaith also mentored a host of interior designers and architects in his practice, including architect Gordon Bunshaft, who started in Snaith's shop as a draftsman and later became president of Skidmore, Owings, and Merrill.

Reese described Snaith as the opposite of Loewy: crude, loud, and rough, although Reese suspected Snaith amplified his roughhewn persona to accentuate the differences between himself and his partner. "Another difference between the two men was that Snaith never minimized the fact that he was Jewish," she wrote. "Loewy, on the other hand, chose to accentuate his Catholic heritage. This was because of the realities of the era, when most company presidents were WASPs. Loewy always recognized the importance of fitting in."

Snaith's first assignment for Loewy was the redesign of a stationery department of a New York department store. Snaith's interiors department eventually morphed into the Department for Specialized Architecture. Joining other departments—Transportation, Product Design, Packaging, and Corporate Identity—Loewy's shop was the first to establish itself as a full-service agency in all design areas.

It is an easy case to make that Snaith and the retail design shop remade the urban and suburban shopping experience. In addition to physical restructuring department stores and other retailers, Loewy and Snaith were among the first consultants to explore the psychology of consumers, tracking their movement through retail space, noting their reactions to décor and lighting, and adapting their findings to retail space on a mass-market scale as well as in individual stores. Snaith recognized also that large, multistory downtown department stores were slowly being replaced by smaller, suburban retail centers. As developers turned the suburban stores into anchors for regional shopping malls, Loewy and Snaith became, for a while, the largest retail and planning consultants in the country. In a company press release, Snaith said, "Based on our research into needs and operations, we developed the modern department store." It is difficult to argue the point.

Bill Snaith understood that big-city retailing required chain stores to move where the customers were. The customers were in the suburbs, at least those newly moneyed, middle-class customers who bought most of the high-end products that Lord & Taylor sold. Snaith and Loewy, one an expert in store design, the other an automotive savant, realized that the shopping experience would be markedly different in the suburbs. Customers came to the shopping center by car and parked away from the main shopping area, which meant that elaborately designed window displays were useless and ineffective as merchandising tools. Instead, the store name was emblazoned high on an exterior wall, visible from entry roads. Instead of the formal, straight counters found in downtown stores, Snaith proposed smaller curved counters placed informally around the floor plan. "The store itself becomes American Suburbia's village green," Loewy wrote. The stockrooms and fitting areas were against the outside walls, the interior selling space within the outer space. Snaith included what Loewy called a "daylight selling area," which was a huge window into the main selling floor. Theoretically, passersby would see the shoppers through the window and be inspired to go inside and spend money. That feature never quite caught on, but many of the other store features were soon to become universal. Lord & Taylor never claimed the new design was a breakthrough, but the chain retained Loewy and Associates for six more branch stores, the last opening in 1959. For Snaith and Loewy, the architectural look of the store was secondary to the shopping experience of the customers.

During World War II, when most companies were suspending major products, Snaith mused on blue-sky ideas for the future of retailing. Snaith completed a study for the Associated Merchandising Corporation that identified four components to big-box retailing. First, department stores were "probably the last of the manually operated large industries." Second, mechanizing retailing would be difficult to accomplish. Third, high labor costs prevented stores from competing against price pressure. Lastly, the operation of downtown department stores was hindered, rather than helped, by the buildings they were housed in.

After the war, Loewy and Associates found the opportunity to put Snaith's theories into practice. The Foley Brothers department store in Houston, Texas, had no outside windows, which allowed Snaith to mechanize at least some of the processes of retailing. The stock-handling system took merchandise from the basement to stockrooms behind the sales floor. The assembly-line aspects of the stock system saved time for the shopper and lowered labor costs. Shortly after the opening of Foley's new store, the firm was asked to institute a similar system for the Gimbel's department store in New York.

Snaith was commissioned to do a similar market study for the Super-market Institute in the early 1960s, and adapted the department store idea of boutiques within a larger store to break up the monotonous rows of similar products in early grocery chains. Most of his theories came directly out of the company's commission in 1947 to revamp the Lucky Supermarket in San Leandro, California. The store featured long aisles with goods arranged by category and brand name. Shoppers used a directory fastened onto a cart to navigate the store. A large display board suggested menus and specials for the day. Shoppers weighed and packaged their produce themselves. The designed San Leandro super-market cost $248,000 (almost $2.8 million today), and company presi-dent Charlie Crouch estimated it would bring in about $39,000 (about $438,000 today) per week. Crouch reported to *Time* magazine that the store grossed $72,000 (about $808,000 today) in its first four weeks.

The consultants were paid $75,000 ($842,000 today) and delivered the report after a full year of research. The report, titled *Super Markets of the Sixties (The Loewy Report to SMI)*, detailed recommendations cadged from Snaith's retailing experience such as bright, cheerful colors and staged lighting. The report's singular breakthrough emphasized that su-permarkets must highlight specialty areas. Another insight recommended that grocery stores expand beyond food and dry goods into stocking greeting cards, books, photo supplies, small appliances, and other items. Anyone who has wandered through a twenty-first-century grocery store knows that most of these suggestions became gospel for grocers.

The Supermarket Institute study polled housewives across every social class. The chief finding from the survey was that most people felt shopping in a supermarket was a mind-numbing experience, as boring as painting walls. Snaith, recalling his days as a theatrical designer, recommended that markets present their wares in special displays that were intended to be dramatic. Shelves and long aisles should be broken up into smaller areas such delis, bakeries, cold boxes, and personal products. Many of Snaith's ideas are still used today, evidenced by the "boutique-style" delis, seafood counters, bakeries, and cheese counters in modern grocery stores.

Snaith's considerable contribution to the reputation of Loewy's company was recognized during his lifetime, as Loewy consistently credited his partner in his memoir and in articles. After Snaith died during open-heart surgery in 1974, Loewy did not emphasize the role his partner had played in building the business in his last book, *Industrial Design*, preferring to erase the designer's legacy. Still, Loewy lists almost two dozen stores and retailers who were Loewy clients, making it clear (by inference, at least) that Snaith's contribution to the success of Loewy and Associates was crucial.

The interior design business also went beyond easily replicated retail spaces. Snaith and other designers tackled railway stations, including the Norfolk and Southern station in Roanoke, Virginia, now experiencing a renovated second life as the Winston O. Link Museum. Today, the train station avoids the monumentality that characterized most urban train stations, using a horizontal elegance to draw the eye to a postmodern columned entrance. Most of the interior was and is spare, with dark woods used in the benches and furniture and a dark terrazzo floor.

From his many commercial retail designs, Snaith also received commissions for transportation interiors. He created Streamline Moderne interiors for the steamships *Panama*, *Cristobal*, and *Acona*. The Panama Line used these small liners to travel to Caribbean ports. In World War II, the *Panama* served as General Dwight Eisenhower's headquarters for the Operation Overlord invasion of Normandy. The interior design was aggressively modernist, and for years a Loewy employee would visit the ships in port to ensure the company's designs were still intact.

In 1958, Loewy and Snaith designed new interiors for the *Atlantic*, a large ocean steamer for the American Banner Line. Loewy, like Henry Dreyfuss and Walter Teague, was one of the first designers to create interiors for airliners. Loewy's work for the Boeing Stratoliner in 1939 and the Lockheed Constellation in 1947 were celebrated in several design case studies. The company's passenger jet designs culminated in his work for Air France's Concorde supersonic airliner.

The firm's most celebrated and influential interior, however, was executed in 1952 for the Lever Building in Manhattan. Built by the architectural firm of Skidmore, Owings, and Merrill, the building was one of the first successful skyscrapers built by American architects in the stark International Style. Considering that International Style designers, who subscribed to the Bauhaus principles of "form follows function" purity of design, were the antithesis of Loewy's aesthetic, the commission was uncharacteristically praised by critics then and now. "I don't know any other building in the city in which so much color has been used with such skill and charm over such a large area," wrote architectural critic Lewis Mumford in the *New Yorker*.

Loewy's Lever Building designers created interiors for each of the subsidiary companies occupying the building. For each company, the design team created and installed unique color schemes for the floor the firm occupied. Each subsidiary color palette complemented the main color scheme used in the building lobby and other common areas.

BY THE END OF WORLD WAR II, Loewy had built his business into one of the major design firms in the country, followed closely by the firms owned by Henry Dreyfuss and Walter Teague. Although industrial design firms grew out of the advertising industry, the way the major designers organized their work provided a footprint for modern consultancies. Instead of myriad salesmen and account executives massaging client egos and entertaining businessmen, Loewy, Teague, and Dreyfuss used senior designers as account managers. Then the company was divided into divisions of specialization, a model still used today by legions of consultant firms ranging from branding companies, Web designers, public relations companies, and some architectural firms.

The use of such a structure to create a consultancy is effective today because the cultural mores of business changed. Most companies don't have the time, inclination, or constitution to be wined and dined by sales reps and account managers. These days, it's better to hear a presentation by the principal leaders and then work directly with a junior representative with the same skill set. It reduces the process to a simple transaction.

The client is missing an important area of business expertise. The consultants are experts in that field and usually provide effective service. It reduces the interaction to, if not Socratic simplicity, then at least an efficiency that would leave the writers of *Mad Men* without a plot. Still, the field of design sprouted quickly, attracting more than two thousand practitioners. Loewy and his colleagues sensed that so many designers hanging out shingles might have diluted their own earning power while lowering the standards of the profession.

So, the Big Three of industrial design decided to establish professional standards. Loewy, Dreyfuss, and Teague invited ten other designers in 1944 to form the Society of Industrial Designers. Loewy became president two years later. One of his first acts was to adopt a code of ethics, later to become a code of practice. The code stated that designers must be honest, loyal to the client, and never work for two clients in direct competition. In addition, designers must not purposely denigrate the business of a colleague, and must refrain from poaching clients and other firms' designers. The society also recommended that designers avoid trying to predict outcomes, never take credit for another's work, never advertise, enter only society-approved contests, and never set up one-person exhibits without the permission of the society. Membership was confined to practitioners or professors of industrial design with "ability, integrity and character." They were required to have designed three or more different products, worked for three different manufacturers, and be sponsored by at least two society members.

The society's leaders were concerned with diluting their influence and, probably more pressingly, losing business to designers that might specialize in design niches. Loewy recalled how the group felt that the society should remain open only to the large design shops, revealing a bit too much about the designers' ulterior motives: "Our small group became appalled and a bit frightened by this stampede led by fast-buck artists."

Jay Doblin, who left the industrial design profession to pursue the academic lifestyle as director of the Institute of Design at the Illinois Institute of Technology and served as president of the Industrial Design Society of America, succinctly characterizes the allure and glamour of the full-service Loewy operation at the height of its powers:

> During the late forties and early fifties, the Loewy office represented
> a truly marvelous design era that will never come again. There was

fun and gaiety, prank-playing and camaraderie amongst us, and the work itself was incredible. We had a hundred major accounts: railroads, automobile companies, bus companies, everybody was a Loewy account. At [age] thirty I was designing for Nabisco, Shell Oil. I had a free pass to go anywhere in the world, hire anybody and build as many models as I wanted. There was always plenty of money. Loewy was running a four million-dollar operation with upwards of two hundred employees, so the place was packed with talented and amusing people. It was a great kick to be part of it and we probably would have worked there for nothing! We had a Loewy band and a Loewy theatrical group, and we were always having wild parties. It was a delirious, crazy, delightful time.

7

Engines of Industry
Tractors, Tour Buses, and Ships

Loewy and his team transformed an entire industry, not in the automotive business, where the designer's influence caught on in earnest after his departure from the production line, but rather in America's fundamental industry—agriculture—when he agreed to take on McCormick-Deering, soon to be known as International Harvester. International Harvester grew out of the 1831 invention by Cyrus McCormick of the mechanized reaper. McCormick's enterprise fought off myriad competitors after forming the company, but the firm finally merged with the Deering Harvester Company in 1902. As more technology came onto the market, McCormick-Deering was in danger of being overtaken by its competitors, all selling some derivation on McCormick's reaper design. Before McCormick's invention, farmers stooped to harvest crops using a sickle or used the more ergonomic scythe to fell crops. At top efficiency, farmers could bring in only about three acres a day. Mechanized agriculture exploded those numbers. In 1902, the company merged with four other equipment manufacturers to form International Harvester. By 1909, it was the fourth-largest corporation in America and the largest farm equipment company in the world.

Loewy's first assignment for the farm equipment giant in 1937 was nearly the same challenge he faced with his first industrial design assignment with Gestetner. The McCormick-Deering cream separator, which used a series of gears and centrifugal plates to separate the heavier cream from fresh milk, was a jumble of metal parts and open gears. The separators mechanized a process that previously occurred as a "natural" system. The machines created extra income for farm families and freed farm wives from the drudgery of manually skimming cream. The last version of the separator before Loewy came on the scene had been designed in 1926.

In the years prior to World War II, most American farmers milked fewer than forty cows, and many kept less than a dozen head to provide milk for a small clientele. Every farm kept a cream separator in frequent use. The large-scale industrial farm where milk was treated and pasteurized at central dairy facilities came much later, in the 1950s and 1960s. The McCormick-Deering separator was efficient, but the exposed gears and multiplaned surfaces were difficult to clean. On the American farm prior to 1945, most of the cleaning of smaller equipment was performed by the women in the household. Farm wives and daughters often were tasked with duties that kept the farm running efficiently while men worked the fields. Loewy used one of his first forays into consumer research to discover how equipment was used and by whom, and how purchasing decisions were made on the farm.

Loewy's redesign of the company's cream separator is a masterpiece of surface makeover. The bulk of the machinery was hidden behind a cream-colored shell complemented by smooth, wide plates that rotated to separate milk into its component parts. The resulting product was perhaps the most streamlined of Loewy's International Harvester designs. As an example of product redesign, it is the longest-lived and most influential of the Harvester product designs.

Loewy also redesigned the company's mainstay Farmall tractor, further influencing the design industry. In the late 1930s, the company hired Loewy and Associates to give International Harvester equipment a distinct modern appearance that would appeal to farmers and their families. Loewy created a smoothly rounded radiator grille design that featured three silver chrome moldings, reminiscent of the Pennsylvania Railroad's speed lines, and a three-dimensional Farmall nameplate.

The Farmall was agriculture's first all-purpose tractor, developed in 1921 to compete with Fordson tractors, manufactured by the Ford Motor Company. From 1917 to 1922, Ford dominated the tractor market, primarily because Ford could manufacture their tractors faster and less expensively than International Harvester, which had older, less adaptable manufacturing plants. The Farmall, created to compete head to head with Ford, could effectively perform all types of light farm work and effectively eliminated the need for horses. It could pull plows, supply power to equipment through the power-takeoff unit, and operate a silo or threshing machine through the use of belt drives. Loewy's redesign of the Farmall returned the company to its former dominance of the agricultural sector.

The Farmall had been designed to farm the new crop cultivars that would initiate the beginning of high-production agriculture. The large, high-set rear wheels could pass over the immature crops, while the narrow front wheels ran between the rows. The company sold its first Farmall to an Iowa farmer in 1924 and the model was still using technology—studded metal wheels, open engine compartments—that predated World War I.

Loewy's mission was to redesign not only the Farmall but also the equipment and parts the company made to complement the versatile tractor. The designer put his research department on the account and found several areas of interest, such as the role farm women played in choosing equipment used on the farm. Loewy and his designers made two major changes to the tractor. They offset the driver's seat and streamlined the engine cowling to make sightlines clear for any operator, whether man, woman, or child. Clear sightlines were crucial to harvesting and cultivating row crops. In addition, the designers studied the somewhat crude seat and created one that was more contoured and ergonomic. Made entirely of metal, the new seat wasn't exactly comfortable, but at least operators could endure a day's plowing. It also required the exhaust stack to be extended so that it could lift the smoke over the driver's head. The new tractor debuted in 1939.

Loewy's great rival, Henry Dreyfuss, also took on an agricultural client: John Deere. Dreyfuss's crucial breakthroughs were redesigns of the radiator and hood components to make maintenance and cleaning easier. He noted that farmers often left the hood off their tractors after working on the engine for the first time and incorporated open space into his redesign to better facilitate engine maintenance. Dreyfuss designed a unified hood and radiator, using a lean-forward shape that suggested straining or pulling a great load. Dreyfuss's other breakthrough design for Deere was the driver's seat. The designer brought in Janet Travell, a specialist in skeletal disorders, to consult on the design. The seat was padded, featured a lumbar support cushion, was adjustable in

This 1936 photograph shows an early version of the International Harvester Farmall tractor, when the company was still known as the McCormick-Deering Company. Loewy streamlined the engine compartment, offset the driver's seat to make planting easier, and later added a three-point wheelbase. The Farmall tractor dominated the agriculture market for decades and inspired John Deere to hire Loewy's rival Henry Dreyfuss to design their tractor fleet. (Raymond Loewy Archive, Courtesy Hagley Museum and Library)

a variety of ways, and far surpassed previous attempts at seat design. Most tractor seats, including Loewy's for International Harvester, used a saddle shape molded to the contours of a generic set of buttocks.

Although the Loewy's equipment designs for the agribusiness sector were influential and responsible for millions of dollars in sales, his lasting contribution came in the creation of one of the most graphically recognizable corporate logos ever designed. The intersected "I" and "H" symbol that Loewy designed replaced a deliberately antique corporate symbol that the company used from 1902 to 1945, featuring an interlocked "I" and "H" set inside a "C".

"I left Chicago for Fort Wayne [Indiana] on the train and sketched a design on a dinner menu. It was reminiscent of the front end of a tractor and its operator." It seems Loewy was forever dashing off memorable logos while traveling on planes and trains. More likely is that Loewy's associates came up with the final product. The logo combines a red lowercase "i" with a black uppercase "H." The enlarged former neatly bisects the latter to create an abstracted version of the Farmall three-point wheel design tractor. The logo design began with block-letter combinations and used a round dot in the "i" to resemble the head of the figure driving. Later incarnations further abstracted the figure, squaring off the dot until the final look was settled. The International Harvester brand, long gone by the turn of the twenty-first century, became so recognized as a symbol of rugged machinery and equipment that the logo was retained when Case, another tractor manufacturer, bought out the venerable company. "The design contradicts [the idea] that trademarks demand thorough, lengthy, expensive research and a great many interviews, tests and polls," Loewy wrote in *Industrial Design*.

The use of red in the logo and Loewy and William Snaith's unerring eye for store design likewise helped inspire Loewy's vision for International Harvester's service centers. The centers needed a distinctive design that would be instantly recognizable, seen easily from a distance or from a rural road or highway, as well as easily replicated on

Loewy and his retail space design partner William Snaith collaborated on the design of International Harvester stores. The modernist glass-and-metal construction allowed customers to see the store's stock, and the bright red pylon (which mimicked Loewy's famous International Harvester logo) that stretched above the roofline acted as a beacon to drivers approaching the stores, most of which were in rural areas. Loewy also designed all the packaging for the company. (Raymond Loewy Archive, Courtesy Hagley Museum and Library)

commercial lots that were rarely uniform in size. Loewy takes credit in *Industrial Design* for creating International Harvester outlets, but his claim to authorship seems unlikely because they were retail spaces— Snaith's bailiwick. The dealership building was modular, so the layout could be expanded, contracted, or realigned on an existing site. In addition, each store featured a rectangular red pylon, the architectural equivalent of the "i" in the company logo, which could be seen at a distance and acted as a brand identifier for farmers and the company's expanding suburban market. The look of the dealerships was replicated across the country, eventually reaching 1,800 stores with standardized design. Coupled with minimalist, color-coordinated packaging for parts and accessories, the International Harvester commissions remade the company's image from a stodgy farm equipment seller into one of the nation's cutting-edge merchandisers.

Standardization of image through architecture, essentially marketing an outlet's "look," became a hallmark of the top-market industrial designers. Examples include Teague's designs for Texaco service stations and Dreyfuss's Bell Telephone public phones. Through color and decorative motifs such as the distinctive red lines atop a Texaco station or the glass-encased telephone booths, these designers created structures that were instantly recognizable at a distance and distinctive amid heavily populated retail areas.

AFTER DESIGNING TRAINS, automobiles, and tractors, Loewy branched out into mass transportation by taking on the Greyhound account. In 1939, the nationally expanding company wanted to standardize its bus fleets. As the company grew, the firm absorbed many local bus lines, making quality control difficult at best. Seeking to solidify its corporate image and standardize its look, Greyhound came to Loewy. His first commission was a cosmetic makeover to give all buses a uniform appearance. He designed a new paint job that unified the entire corporate fleet. He designed a blue-and-white paint scheme that circled each wheel well in a design that recalled ocean waves. Several design historians have noted that the design also recalls the swooping fender shapes of Loewy's 1928 automobile patent.

The standardization of the Greyhound lines was important for the growing company as it sought to become the nation's leading bus carrier. The company was founded in 1914 as a local bus line in Hibbing,

Minnesota, serving iron miners riding from town to dig sites. Its founder, a Hibbing miner named Carl Eric Wickman, originally had sought to open a Hupmobile dealership, but after failing to sell his only model, he converted the large car into a shuttle. By 1918, Wickman had expanded to eighteen buses, and by 1922, he partnered with a bus operation owned by Orville Caesar and bought out several West Coast lines. Early bus lines popped up in cities across the country, all using different vehicles, haphazard schedules, and multiple colors, often directly competing with trolley lines.

Early buses were trucks built to seat about a dozen people in the rear bed (outfitted with seats and an overhead cover). As companies asked coachbuilders to design stretched vehicles that could accommodate seated and standing passengers, riders began calling the elongated vehicles "dachshunds." As might be expected, drivers did not particularly relish the idea of driving a dachshund, so they began to call the new stretch buses "greyhounds." By 1935, the Greyhound Company owned more than 1,700 buses and covered more than 45,000 miles across the Midwest and other areas.

Loewy's commission to standardize the exterior colors segued to creating a new version of the muscular racing dog to replace the company's existing greyhound logo, an animal Loewy referred to as a "fat mongrel." The new logo, which is still in use today (though other designers have tweaked the original), captures the sleek profile of a greyhound in full stride, using simple gray-and-white tones. Orville Caesar, by then Greyhound's chief executive, met with Loewy and asked what he would change first. Loewy immediately parried with the "fat mongrel" remark. "Amazingly, he agreed with me and asked me to do better," Loewy wrote in *Industrial Design*. In this story, Loewy puts in a call to the American Kennel Club and has his staff come up with the logo. It is more likely that a staff designer made the call.

His next assignment for the carrier focused on a new type of bus that would eliminate standing room for riders, introduce more comfortable seats, and include an onboard bathroom. The bus was built by General Motors, although Greyhound had a falling out with the company and its famed chief designer Harley Earl. Loewy accepted the commission in 1940, but production was delayed until after World War II, and the new bus, called the Silversides motor coach, debuted in 1946. For the interior, Loewy created a series of visual safety cues that made long-

Loewy's designs for the Greyhound Bus Company were mostly confined to bus interiors. GM's studios designed the body style. Loewy did redesign the company's canine mascot, slenderizing the "fat mongrel" used previously. (Raymond Loewy Archive, Courtesy Hagley Museum and Library)

distance transportation easier for weary travelers to negotiate. White lines painted on the floor ended at exits with a large white disk featuring a large red arrow pointing directly at exit steps. GM also engineered the bus so it could carry a spare engine.

The company and Loewy adapted the Silversides design into a double-decker version dubbed the Scenicruiser, which was unveiled in 1954 and retired in 1978. The double-decker model was revolutionary for its time. The bus featured an air suspension that transformed long bus trips from bouncing torture into smooth journeys. The bus was air-conditioned and featured a revolving baggage rack. The most visually distinctive design element was the second-deck windshield made of darkened glass, which hid the cattle-car seating within and allowed passengers to unobtrusively observe the outside world. The Scenicruiser's upper observatory level proved popular, and Loewy's designers were so invested in its success (and the possibility of scoring a design victory over GM) that Loewy rented a large, high-ceilinged store—a former powerboat showroom—in New York City at Park Avenue and 45th Street in which to stage a full-size mockup. The Scenicruiser as Loewy and Associates imagined it was a roadway mainstay for decades before the company opted for smaller buses. Plus, Loewy had added one more wheeled design to his vehicular portfolio. *Architectural Forum* famously described Loewy in a 1940 article as being "the only designer in the United States who can cross the country in cars, buses, trains and aircraft that he has designed himself."

As IT TURNED OUT, the designer also could cross the ocean in designs produced at his company. A lifelong nautical enthusiast and owner of several speedboats, Loewy loved ship design but rarely got the chance to put it into practice. His opportunity came in 1933, when a subsidiary of the Pennsylvania Railroad, the Virginia Ferry Corporation, asked him to redesign one of its ferries used in local transportation across the Chesapeake Bay between Norfolk and Cape Charles. Although the Pennsy was celebrated for its extensive rail network, the transportation giant also had its own small navy, which included 342 ferries, tugboats, and barges. The design of the ferry *Princess Anne*, which went into operation in 1936, is considered by most experts—including design historian Arthur J. Pulos—to be the most beautiful and functional streamlined design of the era.

Loewy was fortunate to receive this commission, because there was little opportunity for freelance designers to execute designs for an entire ship. Most ships were designed in shipyards, by largely anonymous shipwrights. Designers often were called on to supply interior decoration and furniture. The rail line wanted Loewy to overhaul the shape of the ferry's exterior and improve interior spaces. Saved from having to create an entire vessel on the drawing board, Loewy and his designers poured their energies into exterior detail and interiors.

The ferry retained certain design tropes that would later be adapted for larger ocean liners. In *Industrial Design*, Loewy claims the *Princess Anne* directly influenced the look of all passenger liners built after it, a bold statement even for Loewy. Others weighed in, calling it the "ferry transport of tomorrow, today." The hull and the upper level and smokestacks were painted cobalt blue separated by vast swaths of white. Another sensuously curving line of blue swept from the prow along the rail line of the deck, giving the ship an illusionary sense of length. Its length was further accentuated by the elongated porthole openings spaced evenly along the promenade deck. "When illuminated at night, the *Princess Anne* looked like a giant liner ready to cross the Atlantic," Loewy wrote. The final design added a dance floor, snack bar, and restaurant,

A drawing of the *Princess Anne* ferry, a commuter ship owned by a Virginia subsidiary of the Pennsylvania Railroad. The final design might be the most perfectly conceived streamlined ship ever built. The hull, upper level, and smokestacks were painted cobalt blue, separated by white. Another curving line of cobalt blue swept from the prow along the rail line of the deck, giving the ship the illusion of length. (Raymond Loewy Archive, Courtesy Hagley Museum and Library)

making the trip on a ferry more like a cruise on an ocean liner. The 260-foot ship was the largest and fastest ship of its category. Loewy's ship designs, although rare, put him ahead of all his contemporaries.

Norman Bel Geddes also created several nautical designs for vast, seagoing streamliners, but as with most of his output, these vehicles existed only on paper. The closest Bel Geddes came to completing an ocean vessel came in his yacht design for Axel Wenner-Gren. Bel Geddes complemented the ship's classic prow lines with a softly rounded bridge resembling an airplane wing. The upper and lower decks also featured the elongated porthole openings favored by Loewy. Some details were plainly impractical, such as a catamaran-style boat launch that jutted from the starboard side, but the design was one of the most achievable of Geddes's many "blue sky" ideas.

Loewy's designs, in collaboration with naval architect George C. Sharp, for the Panama Lines ships in 1936 went against the normative designs for ships. Loewy said as much in a radio interview in 1938. "We are not going to reproduce any classic styles; these ships will be entirely modern." Most American ships featured interiors following classic home styles such as Regency, Queen Anne, or Colonial. Loewy felt that public transportation design should emphasize simplicity. Safety regulations inspired designers to use a host of new materials, such as stainless-steel doors, Formica tabletops set in tubular metal frames, and polished metal banisters, which all "made steamships a machine for living in temporarily." Loewy lessened the harshness of the metallic interiors with carpeting and muted wall colors of coral, blush-beige, and apple green. Loewy later told a magazine writer that industrial design would soon branch out from consumer products into design of transportation hubs, business centers, and airports.

The business of transportation always fascinated Loewy, and his destiny as an industrial designer was converging on the industry where he would make a lasting impression. After a less-than-stellar experience with Hupp Motor Company, Loewy decided to again tackle an automotive client. His work for Studebaker Motor Company would change car design forever.

8

Studebaker Beginnings
Internal Combustion, Internal Dissension, External Design

Raymond Loewy's first experience with car design, the creation of the streamlined 1934 Hupmobile, left the designer unable to control the process once the car went into the manufacturing stage. When Studebaker approached him to design its cars, he was determined to change that outcome. In *Never Leave Well Enough Alone*, in which Loewy's work with Studebaker is barely mentioned (the wound of his struggles on the 1947 car was certainly too fresh), he gives detailed advice on how to control one's work. His hard-earned experience with Hupmobile would convince him to play up his love of American cars and downplay (or at least hide his enthusiasm for) the distinctive cars of his youth in Europe.

The first innovative car designs came out of Europe, particularly France and Germany. The Daimlers and Mercedes that set the automotive standards from 1900 to 1920 would have been seen and studied by a young middle-class boy from Paris. As Loewy wrote in the *Atlantic Monthly* in April 1955, automobile design was less important to the company owners and executives than function and production concerns. "The automobile was an invention, and it looked like one," he wrote.

Automotive journalist Len Frank, editor of *Motor Trend* and *Popular Mechanics* and no fan of Loewy's, described Studebaker's hiring of the famed French designer this way:

> Using his own funds, he designed (in-so-far as we know) and had built an elegant Hupmobile, which he used to sell Hupp on a design

program. A few successful years later, Studebaker bought Loewy, natty pinstriped suit, foulard tie, fresh boutonniere and all. It was the beginning of a peculiar alliance. As the story goes, that handsome Hupmobile was the last automotive task on which Mr. Loewy actually soiled his manicured hands. He was an excellent businessman, an excellent salesman, an excellent chooser of personnel. It was too much to ask him to do the actual design work. It was not too much to ask him to take the credit.

Loewy was prepared to bring one nonnegotiable demand to the table when Studebaker approached Raymond Loewy and Associates in 1938 to be the company's consulting designer. He wanted complete control, and after an initial period where the firm designed the cars in the New York office, Loewy proposed to the Studebaker executives that he create a design office within the massive auto plant complex in South Bend, Indiana. Loewy set the ground rule that company executives, save for project engineers, could not enter the design office unless invited. Many of the in-house car design studios had similar rules. General Motors' legendary stylist Harley Earl kept most of the company executives out of the design studios. Studebaker, which was the fourth-largest automaker in the United States yet well behind in sales to Ford, GM, and Chrysler, was desperate to find a consistent national market. The carmaker agreed to Loewy's conditions. It had little choice.

Loewy described his firm's offices at South Bend in one of the last chapters of *Never Leave Well Enough Alone*. The design department was in Plant 3 and numbered about forty employees. His office was decorated in grays, with black patent leather furniture. He mentions a long window that flooded the office with natural light. Plush offices notwithstanding, Studebaker's position in the pantheon of car manufacturers was anything but preordained. As one of the hundreds of independent car companies at the turn of the century, Studebaker had an advantage over most of its competitors. In 1852, Henry and Clem Studebaker went into business to shoe horses and build and repair wagons, making their first sale of an oak wagon to one George Earl. The price was $175. The Studebakers had come to America in 1736 to join a religious sect in Ephrata, Pennsylvania. The order, the German Baptist Brethren, was an Anabaptist denomination known as the Dunkards (so called from the Anabaptist tradition of baptizing twice, once at birth and again as young adults). The sect held that property should be held in common, and participants should live separated from society and war.

John C. Studebaker and Rebecca Mohler met at an 1820 Dunkard gathering, married, and established a blacksmithing and wagon-building business. The enterprise soon hit rough times as John made a series of bad business loans to members of the Ephrata community. The family moved to Ashland, Ohio. Two sons, Clement and Henry, were encouraged to strike out on their own and settled in South Bend, Indiana. By 1848, John C. Studebaker and his five children also had settled in South Bend. It was the ideal location to start a transportation business, and they decided to again try wagon making. The St. Joseph River provided water and power, as well as transportation from Michigan to Indiana. The railroad lines connected South Bend to another major trade center, Toledo, Ohio. One of John C.'s sons, John Mohler Studebaker, left in 1848 for the California goldfields. He made his fortune, but not panning for gold. Instead, he amassed $8,000 (about $241,000 today) making wheelbarrows in Placerville, California. In addition to being home to Studebaker, this small Gold Rush town featured a retailing hall of fame. The butcher was Philip Armour (whose meatpacking company later hired Raymond Loewy for a packaging makeover); the grocer was Mark Hopkins, a future San Francisco financier; and the dry goods store was owned by Leland Stanford.

John M. returned to South Bend and sank his entire nest egg into the wagon-making business, buying out his brother Henry. He agreed to invest in the family business, although his plans had been to return to California. At the time of his investment, Henry left the precariously financed firm after he was admonished by the pacifist Dunkard elders for building wagons for the military. (The Anabaptist principal of pacifism was transforming into a more practical business ethic.) The brothers built the business slowly, coining a slogan along the way: "Always give a little more than you promise."

John M.'s timing was perfect, and so was the family's decision to focus on building farm wagons as the United States began its westward expansion. Many of the Conestoga wagons—the "covered wagons" familiar to fans of western movies—capable of bringing settlers to the West were made by the Studebakers. Studebaker wagons also were used by farmers, businesses, and the US Army. Studebaker wagons hauled supplies for Ulysses S. Grant's troops at the Battle of the Wilderness and countless other confrontations. Company advertising proclaimed Studebaker to be the leading wagon supplier during the Civil War. The company soon expanded into carriages, which opened into new markets. It was a Studebaker carriage that carried Abraham Lincoln and his wife to Ford's Theater on the night of his assassination, although that fact went unmentioned in its advertising. By 1887, the company's slogan was "The Biggest Vehicle House in the World," and its annual sales were $2 million.

After the Civil War, the Studebakers recognized that westward expansion would mean new markets for their wagons. The completion of the transcontinental railroad meant that the company could efficiently ship completed wagons west and with little competition. The company opened a branch office in San Francisco and soon expanded to New York, Chicago, Omaha, Kansas City, Dallas, and Atlanta—offices that would form the backbone of the company's car dealership network.

In 1897, Frederick Fish, who had married John M. Studebaker's daughter Grace, pushed the family to enter the auto business. Contrary to popular myth, the automotive age did not drive all carriage makers out of business. Some, like Studebaker, adapted their wagon operations to automotive manufacturing. Fish was perhaps the antithesis of the automobile visionary. He was not a tinkerer like Henry Ford, nor was he a swashbuckling salesman like William Durant, founder of General Motors. He was a corporate lawyer who was comfortable in the boardrooms of Wall Street. John M. Studebaker was leery of expansion. The wagon business was booming, and he pointed to the bicycle industry as a negative example of jumping on a transportation trend. Bicycles, after an initial spurt of interest, flattened out as a market. Studebaker tentatively entered the auto market, providing bodies for the New York Electric Vehicle Company and its line of taxis. Eventually, he committed to full auto production in 1902, but the company continued to manufacture wagons.

Fish, of course, was proved right in less than a decade. At the turn of the twentieth century, there were 270 automobile companies operating nationally. By 1904, the gasoline engine had become the standard of the industry, and Studebaker introduced its first the same year, a 16-horsepower, two-cylinder engine. By 1908, there were six hundred automakers vying for buyers. Simultaneously, some of the manufacturers sought mergers or partners to combine companies in order to strengthen a perceived or literal weakness or gain dealership networks. Fish, a dealmaker, approached Garford Motors to propose a merger— only to be rejected by the company because "Studebaker didn't cut much of a figure in the auto business."

Undeterred by the weak endorsement, Fish instead decided in 1910 to buy out another recently merged company, the Everitt-Metzger-Flanders Motor Company. EMF had extensive holdings in the West and South. Another throw-in on the deal was a young mechanic's apprentice earning 15¢ per hour: Harold Vance, destined to become Studebaker's president at the height of the company's success. Shortly after the EMF deal, Studebaker finally mounted a successful takeover of Garford Motors. But the manufacturing advances brought in by Ford and other companies had made all automobiles more affordable. In 1908, 63,000 cars were produced. By 1918, that number had greatly expanded, and the price of a new automobile fell more than 70 percent. Studebaker, after absorbing two car companies, was feeling the pinch. The merger brought in some headstrong executives, including William Metzger, a superlative salesman who rivaled Lee Iacocca as a marketer. One coworker characterized him: "Billy was smart and had the marks of a gentleman. He was cold too, and his smile was that of a Spaniard wiping off a knife." Quite an endorsement for any executive, if not for natives of Spain. The takeover of EMF was costly, and Fish sought extensive financing from Wall Street. The company decided to make a stock offering with investment bankers Goldman Sachs. Thus ownership of the company was shifted from strict family control to a board of directors dominated by finance men.

Within a few years of the merger, the American car market had been fully transformed by Fordism manufacturing. By 1912, half the nation's cars were made by seven carmakers: Ford, GM, Studebaker, Willys-Overland, REO, Hudson, and Packard. By 1916, cars retailing for between $800 and $1,300 represented 51 percent of the market. The multiple mergers that allowed the company to expand and grow eventually in 1911 forced the Studebaker family out of any significant role in the company they founded. J. M. Studebaker remained chairman of the board, but any decision-making power rested in the president and the board— dominated by the financial and banking officials brought in by Fish.

After 1915, Studebaker was overseen by Albert Erskine, company president. Erskine was certainly not a person who grew up on the shop floor. He did have a feel for Studebaker's workforce, giving employees perks and benefits beyond the norm for the time. He was once quoted in a local business magazine as saying, "our men build their very souls into the Studebaker cars." Soulfulness aside, Erskine was a moneyman, an executive more concerned with maintaining a stock price than providing value to the customer above all else. To him, cars were commodities, not products that could be imbued with the dreams and aspirations of the buyer.

Early on, cars were used sparingly—they offered little protection from the weather, and most owners drove them only in warm weather, when roads were not mire-filled morasses. Before cars became ubiquitous, the automobile was perceived, particularly by farmers, as a toy of the rich. Then, as farm work became automated and prices rose, automobile prices dropped as carmakers instituted efficient manufacturing methods. Farmers started buying cars or trucks to get themselves and their goods to market. Urban workers, able to walk to work or ride trolleys, were the last of the working class to adopt cars, holding out until the mid-1920s. The car industry grew almost as fast. In 1904, 13,000 people worked in the automotive industry. By 1919, 651,000 car employees were on the job.

By 1920, innovative work was coming from custom coachbuilders who fashioned one-of-a-kind car bodies for clients ranging from movie stars to business titans. On the less expensive end of the automotive spectrum, Henry Ford, whose famous antimarketing attitude is summed up by his quote "They can have any color they want so long as it's black," was outdone by General Motors, whose visionary leader Alfred Sloan realized that customers would replace serviceable cars if they were inspired by a new color or a slight (and inexpensive) design change. But before Sloan was anointed as its savior, General Motors was created by a hustling salesman named William C. Durant, who owned the Buick Motor Company in Flint, Michigan. Starting in September 1908, Durant went on a buying spree that included Oldsmobile, Cadillac, the Oakland Motor Company (soon renamed Pontiac), and the Rapid Motor Vehicle Company, which eventually became GMC. Durant conceived the idea of an array of brands, models, and colors, but his idea did not initially take off. Durant, more of a wheeler-dealer than a manager, usually was too busy making deals to pay attention to product. Sloan wrote of Durant, "Sometimes I used to feel as if he was always holding a telephone in his hand. I think there were twenty telephones in his private office and a switchboard."

With the company on the brink of bankruptcy after his acquisition spree, the banks financing General Motors ousted Durant. Most of the people who worked for him had a hard time getting him to focus on anything but acquiring more companies. Leo Dunlap, general manager of Oakland Motor Company, said, "when Mr. Durant visited one of his plants, it was like the visitation of a cyclone." After declaring bankruptcy in 1910 as his conglomerate became unmanageable, Durant ceded control of GM to others and founded Chevrolet, creating a car line that competed effectively with Ford's Model T. Durant used his success with Chevrolet as a platform to buy up GM stock, and by 1915, he again held a controlling interest in General Motors. Durant, who had a "my way or the highway" approach, went through several executive officers, including future car company founders Walter Chrysler and Henry Leland (the founder of Cadillac and Lincoln). He also hired Sloan and engineer Charles Kettering. Durant's stock speculation once again brought GM to the brink of bankruptcy, and the company's bankers asked for his resignation, which he supplied on November 30, 1920. Sloan assumed the presidency and became CEO in 1923. Durant eventually left the car business, opening the North Flint Recreation Center, a Michigan bowling alley.

In 1927, Sloan decided to take responsibility for spinning GM style into a separate entity, creating the Art and Color Section of General Motors and hiring Harley Earl to run it. The section had a staff of fifty, and soon the idea of finding a car to suit one's dreams and aspirations became a reality in the car market. Whereas Henry Ford felt that transportation was purely utilitarian, GM executives realized that consumers would pay for a car that fit their desires as their circumstances improved. The Ford Motor Company, whose founder might have been happy to make the Model T for the rest of his life, was forced to shut down the massive River Rouge auto plant and spend $18 million to retool for the manufacture of the sleeker, easier-to-drive Model A. The Model A, not surprisingly, came in different colors.

American automotive design was nondescript until the 1904 Packard introduced a memorable "ox-yoke" grille, which made it immediately identifiable. Until the 1930s, most cars still featured the modular approach to car design: a box containing driver and passengers, a hooded engine compartment, an exposed gas tank, and a trunk. This model was rethought by several designers, most memorably by Amos Northrup in his 1932 Graham Blue Streak. The Graham rounded out the car body and integrated all elements into a coherent whole.

The look of cars was changed forever with the debut of the 1932 Chrysler Airflow. The Big Three automakers were aware of streamlining, but had not made the leap toward incorporating it into car design. Chrysler engineers came closest to getting a streamliner in production, but the company's upper management would not sign off on it. Norman Bel Geddes, in his book *Horizons*, recalls finally convincing Walter Chrysler to go forward with the project. The model was rejected by the public, but the Airflow's instrument panel—which spread the gauges across the dash—was immediately influential. To buyers, the Airflow had little to recommend it beyond its streamlined design. It had little ornamentation. Subsequent poor sales were the result of late delivery problems after the initial product rollout and several serious production defects that gave the model an early reputation as a lemon.

Designer Gordon Buehrig, who would later work for Raymond Loewy, next set the streamline design standard with the 1936 Cord 810, which had no running boards. Most cars at the time needed running boards to help both drivers and passengers gracefully step aboard. The 810 was low enough to eliminate them because it was the first successful American production automobile to use a front-wheel-drive system. It had hidden headlights and air intakes that wrapped around the car's "coffin-nose" front end. Front-wheel drive did not appear in showrooms again until Oldsmobile brought out the 1966 Toronado, because revamping assembly lines to accommodate a new style of engine was prohibitively expensive. Also, rear-wheel drive provides better balance and easier steering. Front-wheel drive eliminates weight and improves gas mileage, two factors that were not a concern in Detroit until the 1970s. The Cord, built by E. L. Cord, who also owned the Auburn and Duesenberg marques, was the 1930s equivalent of today's BMW or Audi—it was superbly engineered and fun to drive (or at least fun for chauffeurs to drive). At $2,500 (about $44,000 today), Cords were more expensive than Cadillacs of the same era. These groundbreaking designs were what Raymond Loewy was up against with the Hupmobile. For Studebaker, the designers had to compete against companies that were better financed, better equipped, and much better at marketing their wares.

Automotive executives found early that making stylish cars was not easy. Independent car companies could not compete on styling changes with GM, Ford, and later Chrysler. Annual styling changes for an entire line of cars was—and still is—prohibitively expensive. In the 1940s, for example, GM spent $35 million (about $450 million today) annually to implement model changes. Smaller car companies either delayed styling changes, spread them out over longer cycles, or went out of business. Car companies needed experts, or at least people with a well-developed sense of aesthetics, to style their new models. Raymond Loewy wrote that "styling, with improved function, would not only sell well but create good will for the company."

In the long run, Loewy was right. His tenets of reduced weight, smaller profiles, and less ornamentation would prove to be visionary, as evidenced by the eventual triumph of the "jellybean" designs of the late twentieth and twenty-first centuries. Still, though Loewy was one of the best-known car stylists of the mid-twentieth century, his automotive styling ideas were long overshadowed by those created by Harley Earl, a flamboyant executive who received his start not on the drawing board but as a custom coachbuilder, adapting existing car bodies to the whims of Hollywood stars and executives. Loewy and Earl remained fierce rivals from the 1930s until Earl's retirement in the late 1950s. The two designers traded jibes, zingers, and criticisms in speeches and in the press—a competitive streak that extended even to their sartorial styles.

Earl was a true Californian. One of his uncles was mayor of Los Angeles. He attended the University of Southern California, where he played football and competed in track, but he dropped out after a year to go to work for his father at the custom coach-building firm Earl Automobile Works. Earl was lured to General Motors in 1927 by Lawrence Fisher and Alfred Sloan, to oversee GM's look and visual marketing. He was eventually responsible for the design of more than 50 million vehicles. Unlike such European-influenced designers as Loewy and other Manhattan-based designers, Earl was not creating looks for a niche of consumers. He was inventing desire for drivers in every state, in every social class, and in every financial strata. Cars were designed, he said, "so every time you get in it, it's a relief—you have a little vacation for a while."

Earl's father did not create one-off cars; he customized existing models. One of Harley's first designs was a $28,000 (just over $457,000 today) touring car constructed in 1918 for silent film comedian Roscoe "Fatty" Arbuckle. A Los Angeles Cadillac dealership subsequently hired the young man to style its clients' cars. His work came to the attention of Lawrence Fisher, GM's president of the Cadillac Division, who hired the young designer to come to Detroit. Detroit was the promised land for car fanatics. Ford Motor Company dominated the industry, but General Motors was growing fast. Henry Ford, the impoverished Michigan farm boy, saw the car as a freedom machine. GM's Sloan and, later, Harley Earl saw the car as a way that Americans could strive for self-improvement.

Earl, once he arrived at GM, was charged with restyling the 1927 LaSalle, which had been conceptualized as a step up in luxury between Buick and Cadillac. He delivered a simple yet elegant design—plainly inspired by the European Hispano-Suiza—that sold more than 50,000 units by 1929. From the start of his career, Earl was seen by his employees and associates as being godlike, viewed best from a distance. His designers feared him but loved his eye for design. Dick Teague, best known for his work on American Motors cars in the 1960s and 1970s, said, "He had charisma in spades. Everyone called him Misterl, all one word. You never called him Harley to his face."

At GM, he instituted many of the same things Loewy later incorporated in his own shop. He switched the design process from a series of sketches to creating three-dimensional clay models. Earl also saved his favorite sketches in a large scrapbook, which he used as a "tickler" file to inspire the design staff. His first assignment after the LaSalle taught him not to outpace the public's expectation of automotive style. The 1929 Buick Silver Anniversary car, dubbed the "pregnant Buick" by industry wags, was a balloony, overinflated body style that the public rejected. Sales in the division dropped 37 percent in a year, a clear indication that Earl's design had overreached. Alfred Sloan recommended that designers take a series of steps in design over the life of a car model, an edict that essentially set the tone for the "planned obsolescence" design model that drove GM for the next six decades. Loewy, in a 1950s article for *Science and Mechanics* magazine, underlined the importance of the evolving design. "Almost every designer responsible for the shape of tomorrow's car has a fairly definite idea of what its appearance will be. That is true because designers are already working on plans for 1954 and 1955. And the intervening models between now and then undoubtedly will incorporate certain changes that will accomplish the ultimate design."

Earl found that the internal politics of GM and Fisher Body made it difficult to bring a design vision to market without significant compromise. He sought to ensure that his design staff avoided being steamrolled in design decisions by other departments. One solution Earl came up with meant that the designers wore white smocks, while the modelers wore dark smocks. The de facto uniforms meant that all GM employees knew who worked for Harley Earl (after several years, Earl eliminated the uniforms).

Earl used a five-year rolling design model that traced gradual changes each year to reach an ultimate vision for a car model. He essentially created the idea of a concept car, with the 1938 Buick "Y-Job." The car looked nothing like what was currently on the road, and it contained many design ideas that would never be incorporated in a GM car until Earl was long gone. It also featured the precursors of the tail fins that would dominate the postwar GM automobiles. The sleek, all-black car had recessed headlamps, a power hood, power windows, a power roof, and wraparound bumpers. It had flush door handles, and it was the first GM car to sport a horizontal grille (a direct steal from the 1938 Lincoln Zephyr). The car also featured a "boat-tail" trunk inspired by earlier "speedsters" such as the Auburn and Cord. The Y-Job (which now resides in the GM Design Center in Warren, Michigan) went on tour all over the country and inspired the GM Motorama, a sort of rolling roadshow extravaganza that debuted the new GM models in cities across the United States.

The disdain that many automotive journalists held for the draftsmanship of Loewy, who rarely drew the designs used for his automotive commissions, could equally apply to Earl. Unable to draw a car beyond a rudimentary side view, he ruled the design studio by reacting to drawings and models, and then suggesting, often profanely, design changes. Typically, he would sit in a director's chair in one of the design studios and communicate his ideas through expletives and sexual metaphors. He created nonsense names like "duflunky" and expected his designers to know what he was talking about. Perhaps his most famous metaphor was "Dagmars," which Earl used to describe the breast-like protuberances used on the heavily chromed bumpers on late 1950s GM luxury cars. Inspired by the buxom television hostess on an early version of NBC's *Tonight Show*, Earl used these inadequate descriptions to direct designers in adapting designs. According to William Mitchell, Earl's successor as head of GM design,

> When he wanted something it would be very difficult to work with
> him because he knew what he wanted and he couldn't draw it for

you. And he would be so impatient with you if you didn't get it. He'd just be furious if you didn't get what the hell he was telling you. He had a chair like a director in a studio in Hollywood. He came from Hollywood. He would sit in that chair, and the way he would do it, he would sit and have all the people around him . . . and everyone would run around like a bunch of monkeys.

Earl was the most powerful car designer ever to oversee an automotive product line. He was on a first-name basis with Alfred Sloan, and knew most of the car men of the era. Still, he had to continually prove himself to the engineering and manufacturing executives. Like Loewy with his urbane Euro-persona, Earl had to create an image for himself not only in the car industry but also in his own company. Frank Hershey, GM's chief car designer in the 1930s and '40s, recalled his boss this way: "Harley was always so image-conscious. Detroit was so macho in those days. Everything was macho—the Dodge brothers, the Fisher brothers, all those people. And they used to make fun of Harley's neckties and suits and shirts. All the time. The big shots at GM couldn't understand these either, but that was part of Harley Earl's image. He was showing them he was a designer by wearing all this stuff. He paid twenty-five dollars for a shirt. He was a big man, he should have dressed conservatively, but he didn't. He had an image. He had to sell himself."

Dazzling wardrobe notwithstanding, most of his employees were afraid of him. Yet, like Loewy, he also was shy when faced with unfamiliar social situations. He preferred to talk to the chief designer in each studio, often addressing all comments on a design only to the chief, even if the object of the comments happened to be in the room. Bill Mitchell even took time to explain to Earl that his manner intimidated staffers, but Earl never adapted his methods to be more collegial. Instead, he would often roam the design studios after hours, making observations and then communicating any changes to the chief designers. He was a workaholic, often working from seven in the morning through nine at night. Like Loewy, Earl preferred to be the final arbiter of a design. Kenneth Coppock, chief Chevrolet designer in the 1930s, explained Earl's talent. "Earl wasn't a designer himself. But he was one of the finest critics of design ever to come along."

Earl's design aesthetic, which favored large, low-slung cars with pillowy body molding, reflected his personality. As designer Gene Garfinkle recalled, "He was a very strong, big man and his sense of form had that—everything was like an overstuffed couch." He separated the design teams for Chevrolet, Pontiac, Oldsmobile, Buick, and Cadillac and forbade collaboration between the different studios. A designer could be summarily fired just for talking to another designer or for stepping into a competing studio. His design staff called him "the Shadow" because he loomed over his employees. A bulletin-board sentiment that made the rounds at GM read, "Our father who art in styling, Harley be thy name."

At General Motors, where the flashiest executive wore a club tie to accompany his white shirt and blue suit, Earl was resplendent. He wore custom-made shirts in colors ranging from salmon to azure to puce. Legend had it that he always bought two suits when shopping, keeping one at home and storing the other at his office. At noon, he would change into the "office suit," preserving the razor-sharp creases in his trousers. More than one designer felt that Loewy and Earl were cut from the same cloth. Car historian Edson Armi wrote, "They led the life of jet-setters and designed cars with the certainty and abandon of demigods."

Like Loewy, Earl made it clear that all credit for any design emerging from the studios would go to him. GM designer Strother McMinn, later an instructor at the automotive design school the Art Center and College of Design, best described Earl's design style: "Harley Earl designed a car so that when you walked around it, you'd be entertained the whole trip." Loewy's take? "Big companies jumped on the 'style' bandwagon, but what so-called functional factor did they select? Bulk—a sorry choice. Bulk gets to be habit-forming, and bulk means weight. To this manufacturers added 'flash.' So they got into a spiral of increased weight and ornamentation. This led to the horsepower rat-race and the chrome gadget rat-race—a costly combination."

THE RAT RACE FOR STUDEBAKER executives centered on producing a consistent car. They were not much concerned about what it looked like—the bottom line was whether it sold. One remedy for Studebaker's production problems was the construction of the Studebaker Proving Grounds. The 840-acre facility, built in 1926, was the first of its kind in the country. It featured a three-mile banked test track with every manner of road surface built into it. The company even planted a massive stand of pine trees that spelled out "Studebaker." It is still visible from the air today, decades after the demise of the company.

The Studebaker Proving Ground provided the company a private area to test designs. The track featured areas where test drivers could simulate a variety of driving conditions. The site also has the world's largest living sign, made of trees planted in the late 1930s. It remains visible today as part of the Bendix Woods County Park in St. Joseph County, Indiana. (From the Collection of the Studebaker National Museum, South Bend, Indiana)

Despite the investment in infrastructure, Studebaker struggled to make consistent profits and steady growth. Albert Erskine's lack of heft as a car man—he was more likely to make decisions based on financial efficiency than on marketing dazzling cars—would later be repeated at many different automakers, particularly by corporate "bean counters" such as the Ford "Whiz Kids" hires like Robert McNamara, Arjay Miller, and Charles "Tex" Thornton.

Erskine was born in Huntsville, Alabama, and entered the business world early as a 16-year-old office clerk for the Mobile and Ohio Railroad. He worked his way up through positions at the American Cotton Company and Underwood Typewriter, and he joined Studebaker in 1911 as treasurer. As president, Erskine introduced a line of Studebaker cars at different price points: the Four, the Light Six, and the Big Six. A 1921 entreaty to merge with Maxwell Motor (remembered today for its role as a reluctant-to-start car used by the purportedly miserly comedian Jack Benny on his radio show) met with rejection. Maxwell had a brilliant engineer on its staff, Walter Chrysler, who would introduce a revolutionary high-compression engine shortly after the Studebaker offer was turned down. The Chrysler engine convinced bankers to allow Maxwell to expand, a move that led to the eventual formation of Chrysler Corporation. The high-compression engine paved the way for cars to obtain power to run more systems and drive more efficiently at faster speeds.

Meanwhile, the state-of-the-art manufacturing plant that Erskine authorized put the company in the running as a major player in the auto industry. The plant, featuring 7.5 million feet of covered floor space on 126 acres, was built in South Bend after some debate over establishing a presence in Detroit, a decision that eventually would haunt the company.

That the company did not site its manufacturing plant in Detroit or at least near the Michigan manufacturing center was a strategic error. Although most of the independent car companies based in Detroit are gone today, the nameplates are memorable: Packard, Hudson, Hupmobile, Graham-Paige, Maxwell, and Chalmers. Some survived by merging, such as Nash-Kelvinator merging with Hudson to form American Motors. Eventually, none would be able to compete with the Big Three, and most of these manufacturing plants were torn down or abandoned. The transportation and marketing networks were set up with Detroit as its locus, and those outside its influence struggled.

Erskine made two of the firm's most important hires in 1926. He realized the company needed executives who cared more about cars than finance. Harold Vance, who began his working career as a mechanics apprentice for EMF, came in to head the manufacturing department. Paul Hoffman, a sales specialist who earned millions at his Southern California car dealership, energized Studebaker's sales force. Hoffman, who later headed the Ford Foundation in 1950 and became delegate to the United Nations in 1956, was the perfect candidate to jazz up a lackluster sales force. He made his way to California peddling a line of auto accessories, and he won an award for his essay "How to Sell a Studebaker Car."

A man coming into the sales room should always be met with the idea that he has come in to buy a car that day. That attitude will make the salesman thoroughly alert, courteous, and will bring a smile to his face, the smile that always comes when you have picked a live one. A smiling, courteous salesman establishes a receptive mood on the part of the prospect. That mood is a short cut to the information that is necessary if the salesman wants to fit his talk to meet the particular case in hand. Once the prospect has started talking, it is comparatively easy to find out whether he can buy a car today, this week, next month, or next year, also whether he has a horse and buggy, diamonds, real estate, mining stock, trust deed, mortgage or an equity in a pipe dream as part payment on the car . . . The man who comes into the showroom with only an equity in a pipe dream often knows people with real money, and courtesy to him will, in many a case, uncover the trail of a real live one.

Erskine started the 1920s by asking Studebaker engineers to design a small family car to compete with the Ford and Chevrolet entry-level cars. Underpowered, the Erskine, one of the first in a long line of tin-eared model names, was a failure shortly after introduction. Those cars were joined by "the Sixes," called President, Commander, and, unfortunately, the Dictator (later changed to the Director), no doubt named by an admirer of European fascists. The company ended the decade by purchasing the Pierce-Arrow luxury brand; it would turn into a draining purchase as the nation headed into the Great Depression. Erskine mirrored contemporary business leaders of the late 1920s by neglecting the product to focus on maintaining the Wall Street stock price. The Depression first eviscerated the luxury car market. Luxury auto sales fell from 150,000 in 1929 to 10,000 in 1937. Amid such uncertainty, Erskine embarked on a program to pay large dividends to stockholders. The company paid out $2.6 million when it had produced little or no profit. In the meantime, car companies were falling by the wayside, much like discarded tires. Carmakers Graham-Paige, Durant, REO, Auburn, Hupp, Franklin, Peerless, and Jordan all were out of business by 1935. Auto sales nationally slalomed like a distracted driver, dropping from 1929 to 1932, rising in 1935, and decreasing from 1937 to 1938.

As automakers sought to boost sales, they began to incorporate more features into their cars, features that previously were available only in luxury makes. GM lowered costs and prices by introducing standardized outer body styles. The GM "A" body was used for Chevrolet and Pontiac. The "B" body was used for Buick, Oldsmobile, and LaSalle, and the "C" body was used for Buick, Oldsmobile, and Cadillac. Simultaneously, in South Bend in 1931, Erskine decided the Depression was over and authorized another expansion.

Erskine took a meeting with two engineers, Ralph Vail and Roy Cole, who had engineered a prototype sedan for Willys-Overland. The company could not afford to put it into production, so Erskine bought the prototype for Studebaker and then brought Vail and Cole into the company. He called the new car the Rockne, named for South Bend's most famous citizen, University of Notre Dame football coach Knute Rockne. He sent Harold Vance to create a manufacturing facility for the new car in Detroit and to oversee the introduction of the Rockne. The legendary coach died in a plane crash in 1931 before the car debuted, which made the new model difficult to market, to say the least. Erskine issued even more dividends to stockholders shortly after the Rockne fizzled. To be fair to Erskine, most car companies issued dividends, but most did it profitably. Paul Hoffman, who would also rise to the presidency of Studebaker, recalled, "[We were going to] bring out the Rockne, which was our challenge to Ford. But Mr. Erskine was a year too early. He made the awful mistake of expanding in a dying market." Starved for cash by Erskine's ill-advised stock-support program, Studebaker went into receivership in 1933. The factory shut its doors.

The Great Depression devastated Studebaker's sales, which declined by 62 percent. Still, Studebaker fared better than GM, whose sales fell by 71 percent. Erskine's decision to issue dividends and spend money on the Rockne meant there were few cash reserves to cover losses, making it impossible for the company to recover from its debts. After presiding over the bankruptcy, Erskine committed suicide in 1933. Hoffman and Vance took over and moved the company out of bankruptcy in 1935. They consolidated all operations, including the Rockne plant, to South Bend and closed a few plants. They cut costs relentlessly, unloaded poor performers such as Pierce Arrow and White Motor, and pulled the company into the black. It was the first car company to emerge from receivership in the history of the United States. One of Hoffman and Vance's first acts as top executives was to approve production for a new model, the 1939 Champion. The company named Roy Cole chief engineer, and he set the new line featuring the Dictator, Commander, and President models. The company's advertising called the new models "Champions for 1935."

ENTER RAYMOND LOEWY. The independent designer came to Studebaker in 1936 because, as Vance said, "we felt that our own designers were paying too much attention to the production engineers instead of letting themselves go." There was no one more ready to ignore production engineers and pursue his own vision than Loewy. Working several model years in advance, in line with standard practice at car companies, Loewy started work on the 1938 product line, including what was to be the groundbreaking 1939 Champion. The new improvements for 1938 from the Loewy shop were immediately noticed. The previously pedestrian hoods were sleekly slanted, and the designers integrated the headlights into the front fenders. Loewy showed no hesitation in bringing in any person who could improve the product. He hired several women to help design interiors for the Studebaker product line. Helen Dryden was a contract hire who worked on Studebaker interiors for the 1938–40 models. Later, he would hire other women to work on car accounts, including Virginia Spence and her daughter Nancy, and Audrey Moore Hodges, all of whom worked in the clay modeling studio. Although few women were allowed in main design studios, GM had seven distaff designers working primarily on interiors and instrument panels. One female GM designer, Peggy Sauer, left GM to work for Loewy's studio and ended up working on the interiors for the Avanti.

Audrey Moore Hodges recalled that her work was never judged as inferior because of her gender. In fact, the Loewy system ensured that every designer's work would be judged equally. "Each one of us was given a number for our designs. They didn't go by name; we were given a number, so no one ever knew whose work they were judging." Hodges further recalled that Loewy "was a very fine person to work with, and he was so chic that you just felt everything had to be super chic in doing things in styling, and you knew he expected the cleanest and the crispest, because that's the kind of presentation he made."

As he did at Hupmobile, Loewy and his design team started its tenure by designing discreet but different cosmetic adjustments to existing models. The addition of such Loewy trademarks as speed lines and a gracefully curving split grille earned the 1939 Studebaker President an award as "best-looking car of the year" by the American Federation of

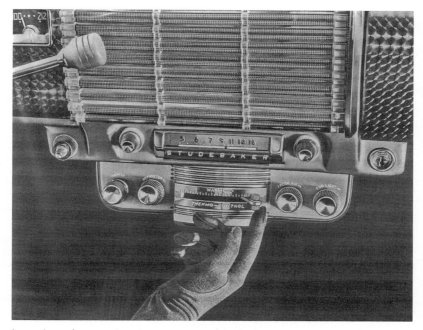

Loewy's work on car interiors is not as celebrated as his more innovative work on body styling, but he employed some of the industry's first female designers to create interiors and unique control areas for drivers. (Raymond Loewy Archive, Courtesy Hagley Museum and Library)

the Arts. Loewy convinced the executives and engineers that the company should differentiate itself by building lighter, more fuel-efficient cars. He directed the design team to plaster signs throughout the design shop. The message? "Weight is the Enemy."

It was in the late 1930s that Loewy identified a consumer desire that would not fully mature until the mid-1970s. The smaller, sleeker car profile, influenced by Loewy's European roots, was truly ahead of its time. Loewy would never be vindicated in his belief that Americans would opt for sportier, fuel-efficient cars until nearly the end of his life, when BMW's 2002 and other models created the sport sedan category that dominates today's car market. "I tried hard to convince management, as early as 1939, that there existed among millions of Americans a segment, profitable to Studebaker, that could not find the kind of car they wanted to buy among the Big Three offerings," said Loewy in an interview. "What these buyers want is a slender, compact automobile."

Studebakers weighed about 15 percent less than comparable Fords or Chevrolets, resulting in markedly better gas mileage and less wear and tear on components. Instead of adding on chrome, Loewy gave the Commander and President models two-toned paint jobs. Inspired by the swooping white teardrop shapes he created for Greyhound buses, Loewy set off his complementary paint schemes with understated strips of chrome. The cars produced by Loewy's shop during the war years under the direction of Virgil Exner incorporated some design tropes from GM, where Exner had worked previously. Exner created an integrated trunk that was inspired by the Cadillac Sixty Special and used wide, horizontal grilles inspired by Pontiacs and the Lincoln Zephyr. Finally, the company acquiesced to eliminating heavy decoration on all vehicles. "The astonishingly spare use of brightwork was one feature that stands out and anticipated the 'less is more' school of automotive design [by] two decades," wrote one automotive historian.

By the end of the 1930s, the wide range of style choices and luxurious appointments were beginning to be integrated into makes and models that were previously stripped-down models. A 1939 Buick or Studebaker

The 1939 Commander was the first acclaimed design that Loewy executed for Studebaker. The company liked the work well enough to hire his company to design all their cars. Virgil Exner, who would run afoul of Loewy's indifferent management style, when he led the design team for the 1947 Studebaker, was the chief designer for this model. (Raymond Loewy Archive, Courtesy Hagley Museum and Library)

offered as standard equipment luxury options that would have been available only on such makes as Packard, Cadillac, and Pierce-Arrow before the Depression. Though money was still tight at the end of the 1930s, the consumer market was ready to assert a formidable purchasing power that had been dormant since 1929. Most families in the country were used to doing without throughout the 1930s. Families stopped buying major ticket items and started gardening and canning their own food, and sewing and mending their own clothes. Most homeowners tried to do their own home repairs. Birth rates declined, as did the number of divorces.

The three major car manufacturers brought out striking new designs in 1938 and 1939. The new 1939 Studebaker came in about 650 to 800 pounds lighter than any competing model from the Big Three. Gas mileage for the new model of the Champion improved by 10 to 25 percent, depending on the trim version. The design emphasized the chrome grille and streamlined body, while buyers reacted favorably to the model's climate control system. The Champion was sold to 72,000 drivers its first year, and by 1940, Studebaker had moved up to eighth in national sales, overtaking Hudson. The company hired 1,500 new workers to handle the increased production.

The United States' entrance into World War II in 1941 led to a shutdown of new car design at all auto companies in order to concentrate on war production. When the war broke out, the plant switched to war materiel. Studebaker worked itself out of the remnants of its Depression-related financial difficulty by ramping up war production. By the end of the war, the company would produce 63,000 engines used in the Boeing B-17 Flying Fortress and 200,000 heavy trucks, most of which went to Russia. (For years afterward, the word "Studebaker" meant "truck" in Russian slang.) An amphibious assault vehicle built in Studebaker facilities, the Weasel, was used on beachheads in Normandy, Italy, and the Pacific theater. The outbreak of war had zoomed the company into profitability during the Civil War and both world wars. Considering the pacifist nature of the Studebaker family, the company's dependence on war materiel manufacture to increase profits held some irony. With the national economy steadily moving upward, consumers again felt comfortable making larger purchases, including cars.

After the debut of the 1939 models, the ramp-up to the war and improving economy allowed Loewy to put together what would amount to an all-star team of designers. Although the 1939 Champion design was accomplished in Studebaker's Rockne plant in Detroit (the Rockne was the company's luxury sedan), the wartime redesign was to be located within the Studebaker plant in South Bend, Indiana. Loewy designated Virgil Exner, a talented designer who was a native of Buchanan, Indiana, near South Bend, to run the in-house studio. The designers in the Loewy studio could work on the entire range of car design jobs, and the gift of freedom from specialization made them better artists. Bob Andrews, one of the designers of the Avanti, described the policy. "We designed in three dimensions there. There was a difference between Raymond Loewy and General Motors, Ford and the rest of them . . . In simple form, being with Raymond Loewy was like being a member of the New York Yankees. I enjoyed the input of our top marketing people. Mr. Loewy never overused marketing . . . he never became a slave of it but used it as an absolutely viable tool . . . We had Nabisco, Shell Oil and all these different top company accounts, and you'd get tuned in."

Virgil Exner's early life was quite different than Loewy's. He was born out of wedlock in Ann Arbor, Michigan, and was adopted as a baby by George and Iva Exner. George Exner was a machinist, and the youngster picked up an interest in mechanics through him. Virgil had attended Notre Dame for several years, but left when the art professors there urged him to enter the professional ranks. He caught on with Advertising Artists Inc., a South Bend agency that handled Studebaker's advertising. He drew print ads and a few blue-sky car drawings. Inspired by a friend's suggestion, he asked to meet with Harley Earl, and the legendarily difficult executive hired him on the spot. Shortly thereafter, Earl promoted him to head of the Pontiac design studio. At age 24, he was the youngest design studio executive at GM. After three years, Loewy lured Exner back to South Bend. Exner commuted to South Bend every four weeks, staying for a week to service the account.

"He has so many ideas, he didn't know when to turn them off," designer Robert Bourke said of Exner. "In other words, he would throw [so] many ideas, 'hot licks,' and various body configurations into a machine."

Meanwhile, the War Production Board set quotas to preserve war materials and to help auto companies switch to war production. In addition, the board sought to even the postwar playing field by decreeing that manufacturers could not "use human or technological resources to develop postwar models." The rule applied to all design studios in the

The Studebaker design brain trust in the late 1940s. *From left*: A. Baker "Barney" Barnhart, Raymond Loewy, Dorothy (last name unknown), Gordon Buehrig (acclaimed designer of the Cord and Auburn classic cars), and Robert Bourke. In the background, note the framed prints of the *Princess Anne* ferry and Greyhound Scenicruiser prototype design. (From the Collection of the Studebaker National Museum, South Bend, Indiana)

automotive industry that were owned by the company. Loewy was the owner of Studebaker's studio, and the carmaker was under no obligation to ask him to stop designing postwar cars. And so Loewy's group began work on postwar designs in 1942 while the plant produced troop-carrier trucks for the war effort. By contrast, when Ford Motor Company designers attempted to secretly run a skunkworks operation during the early years of the war, Ford executives unceremoniously shut it down.

ALTHOUGH AUTOMOBILE DESIGN was the clearest measure of consumer responsiveness for most of America, the design critics who began to write seriously about art and design in the 1930s rarely praised American auto design. New York's Museum of Modern Art (MOMA) in 1951 mounted one of the first art exhibits devoted solely to car design but did not include an American car designed after 1938. The design critics, many of them subscribing to the Bauhaus theory of "form follows function," theorized that American cars failed as good design because the form had nothing to do with function. Everything about American cars was there on the surface. To paraphrase Duke Ellington's aphorism about music, for American car buyers, if it looked good, it was good. Alfred Barr, director of the MOMA, wrote of the role of the artist in machine art: "He does not embellish or elaborate, but refines, simplifies and perfects." To which Earl or Exner might have replied, "Not in Detroit, pal."

Bill Mitchell, who took over Earl's fiefdom in the late 1950s, went even further in an interview with Edson Armi, lumping the purist art critics into a stereotype of snobbish urban aesthetes who rarely deigned to ride in an automobile. "Those cliff dwellers didn't like cars. The inspiration for automobiles never came from New York. [Consumer advocate Ralph] Nader is a cliff dweller—he walks to work. I could give him a ride and give him a stroke."

The essential design trope for 1930s car designers was the teardrop, which came close to the optimal design shape sought by the Bauhaus design school. But aside from interest in Buckminster Fuller's rounded Dymaxion car, the nation's car-buying public was indifferent to perfect shapes, teardrop or otherwise. As the dealers might have said, pure form does not move the metal. GM's Earl said people make the decision to buy a car in the street, not in the showroom. And so designers consciously created their cars to look monumental against the sky and horizon.

The American car industry was self-reverential. Most companies ignored what other automakers overseas were doing. One of the first international designs to influence American car designers was the Italian Cistalia, designed by Pininfarina. The Cistalia was the star of the 1951 MOMA exhibition. *Industrial Design* magazine wrote that the Cistalia was the epitome of car design and singled out Loewy for praise as the first car designer to fold the European influences into an American design. "Loewy lowered his lance against two All-American fetishes: conspicuous consumption [especially chrome] and the Big Package. The public will soon be ready for a car reflecting European design trends." Unfortunately, the writer was a few decades off. Loewy said later about his time at Studebaker, "One of my main and generally little known contributions to Studebaker was my frequent presence in Europe, and a conviction that a lot could learned there about automobiles—especially about suspension, roadability and lower gas consumption."

At the time of Loewy's hire, Studebaker had been experiencing sales years of feast or famine since 1929. Sales plummeted from 1929 to 1932, picked up in 1935, and decreased in 1937 and 1938. The Big Three controlled 90 percent of the market, and the independent automakers cut up the other 10 percent. Loewy's design team took its first major assignment and reinforced his mantra to the designers: "Weight is the enemy; Whatever saves weight saves cost; The car must look fast, whether in motion or stationary." Loewy also had a healthy disdain for working in concert with the engineering department, preferring them to fit the mechanics to the design. "There is much to be gained working backward from optimal form to mechanics," he would say, using movie star Betty Grable as an example. "Her liver and kidneys are no doubt adorable, [but] I would rather have her with skin than without."

Loewy asked Exner, an Indiana native, to move back to South Bend in 1941. Exner supervised the shop, but Loewy reserved the right to approve all designs, a decision that had the potential to cause delays and backlogs. In South Bend, Exner hired Robert Bourke and Frank Ahlroth as his design staff. Bourke started his career at Sears, "designing such memorable devices as manure spreaders, power tools, outboard motors, washing machines and refrigerators. I had always wanted to design automobiles." Loewy, in grandiose fashion, added to the design team in 1942 by hiring Gordon Buehrig, an already legendary designer who created the memorably streamlined chassis for the Cord 810 and the Auburn Speedster. Like so many automotive designers, Buehrig worked for a succession of automakers. He started as an apprentice at Gotfredson Body Shop in Detroit and learned how to visualize rolling sculpture, which would ultimately inspire the title for his 1975 memoir.

Buehrig worked for relatively small companies, such as LeBaron, Briggs, and Packard. He made it to Harley Earl's GM studios, but soon left to join the Stutz Motor Car Company. "I was a very naïve young man," he wrote, "I quit a few months after joining [GM] to take the job . . . I left the strongest company to work for a small company which was in financial trouble." He then moved again to create Duesenberg one-off models. Buehrig found his automotive creativity supercharged by working on the "Doozies," which were highly sought-after luxury cars that could be customized for each purchaser. Buehrig also created the "bird-in-flight" abstracted hood ornament known as the "Duesen-bird."

As this automotive all-star design team came together, the tastes of postwar car buyers were on the brink of change. Even before the war, appliance magnate Powell Crosley created a small two-seater coupe

The Loewy Studebaker studio in South Bend, around 1945, at the height of its success. *Left to right*: Holden "Bob" Koto, Larry Brown (*far rear*), Virginia Spence (*black dress*), Vince Gardner, Robert Bourke (*facing camera in necktie*), Nancy Spence, Frank Ahlroth (*wearing vest*), Jack Aldrich, John Reinhart (*visible suspenders*), Virgil Exner, John Bird, Raymond Loewy, and Gordon Buehrig (*seated on stool*). (From the Collection of the Studebaker National Museum, South Bend, Indiana)

that was manufactured from other companies' parts on the Crosley assembly line. Although the car was simple—it featured sliding windows and hand-operated windshield wipers—it found a market during the war, where used models sold for more than the $325 original price tag. The Big Three's postwar models, however, did not address the market for smaller, more affordable cars. All major manufacturers followed the Harley Earl "balloon" style, overladen with chrome and powered by big engines. The majors did start a small car program in response to customer pressure, but cancelled it in 1948, pleading lack of resources and manufacturing space.

GM's postwar cars were epitomized by Earl's seminal design, the 1948 Cadillac. The long, sleek car featured emerging "fins" that were directly inspired by the Lockheed P-38 twin-fuselage fighter plane. The initial Cadillac fins were subtle and worked harmoniously within the larger design, a quality that was soon to change as designers and consumers began to demand more aircraft-related features. Soon fins would grow larger, and car hoods would be peppered with air intakes, gunsight or- namentation, and faux propellers, or "spinners."

The Studebaker team kept recruiting major talent to add to its roster. Buehrig, during periods when he oversaw the shop, also hired Vince Gardner, Frank Aldrich, John Reinhart, who went on to create the 1956 Lincoln Continental Mark II, and Holden "Bob" Koto. As is the case with most "dream teams," Loewy's cast of designers had trouble working together almost from the start. Buehrig was originally brought in to run the shop, but Loewy reversed that decision and promoted Exner. Both men received conflicting messages from various executives regarding who was in charge.

Buehrig drily noted the less-than-optimum working conditions in an interview. "Virg had been there before I was, and Virg was my assistant, and then Loewy came out from New York one day and got mad at me and made Exner manager and made me his assistant. Then in another period he came out and fired Exner and put me in charge. I mean, it was a funny political situation." Apparently. In day-to-day assignments, Buehrig and Exner got along well; most of the resentment came from Loewy's laissez-faire attitude, frequent extended absences, and insis- tence on claiming credit for group designs. Exner, "a man of immense ego," was also having trouble ceding design credit to Loewy. Although Loewy would travel to South Bend to give major presentations, the bulk of the design work was accomplished by the in-house team.

Bob Bourke, whose lesser status on the team gave him some dis- tance from the controversy, recalls the atmosphere in the design shop as tense. "Loewy would show up out there from time to time, and Ex would mumble under his breath. Ex could get mad at somebody and stay mad forever . . . One thing he didn't understand is the fact that everything in this world has to be promoted or sold in some way or another, and it's not the guy that really does the [design]. It's awfully important to have somebody who gives that guy the opportunity to design something."

The years before and during the war required "the guy that gave designers the opportunity to design something" to crank up his sales- manship to keep commissions coming in. Bourke said of Loewy, "I'd always admired him—especially his ability to sell advanced designs to recalcitrant executives." But Bourke also appreciated Exner's point of view. "Ex felt a man was either a designer or a promoter." For Exner, his boss occupied the latter category. Buehrig, at least in his public writings, said he had little trouble working with Exner or Loewy. "We had mutual respect for each other and managed the office as a team."

Thus the Studebaker designs of 1939 to 1948 were not heavily influenced by Loewy's discerning eye. Unfortunately, Loewy's infrequent stops in South Bend also caused problems with Studebaker executives. In fact, Loewy's European charm and cultured ways were lost on most Studebaker executives, particularly Roy Cole, head of Studebaker's engineering department. Although Loewy could talk about car design with executives and mechanics, he was not what Detroit executives refer to as a "car guy." Car guys were executives who lived and breathed automotive enthusiasm. They loved buying cars, working on cars, and building cars. Loewy, with his precise mustache, continental clothes, and cologne, was not just "a Frog"; to these businessmen, it was as if he were an unapproachable alien.

Bill Mitchell defined the culture of the car guy. "You have to know them all—the histories. And if you are an icebox designer—fine. Damn few people can do a car!" Asked what car designers know about iceboxes, Mitchell continued, "They don't give a damn about it! They are different people entirely. That's why Raymond Loewy was not a car designer; he wanted to be."

By 1942, the draft and enlistments had winnowed the staff down to Exner, Bourke, and modeler Frank Ahlroth. The 1939 Champion had set the style for Studebakers to come, and Exner and Loewy introduced flush-mounted headlights, which would become a Studebaker trademark, and an extended hood rounded in a "prow" that also defined the auto line until the 1960s. For 1941, Loewy reworked the entire Studebaker line into a new body style for the Commander and President models, titled Land Cruiser. Loewy and Exner created framed, complementary color schemes separated by thin lines of chrome. The redesigns were successful, yet the sales figures were not conclusive, thanks in large part to the war economy. The critical advantage Studebaker gained by creating an independent design studio would bear fruit for the 1947 redesign. The rest of the auto industry, hamstrung by the war production limitations, rolled out designs from 1942 that had been little changed with just cosmetic improvements. Studebaker was the only company to have entirely new designs ready for 1947. Reaching that goal would be anything but easy.

Virgil Exner (*left*) and two unidentified Studebaker employees look over the 1949 Champion. Exner, who fought Loewy's hands-off management style and credit-grabbing policies, emerged as the company's top design executive after production chief Roy Cole secretly asked Exner to design the 1947 postwar model while Loewy designers also worked on the design. (From the Collection of the Studebaker National Museum, South Bend, Indiana)

Paul Hoffman and Roy Cole wanted Exner to get a head start on the postwar collection and encouraged him to set up a design studio in his house. "I agreed to start immediately," Exner said in an interview. "I cleaned out one of our bedrooms at home . . . This was my job then, to be worked on nights and weekends."

"First by far with a postwar car," was the tagline used in the company's 1947 advertising. The tensions between the designers and the often-absent Loewy, coupled with tensions between company executives and Loewy, resulted in corporate intrigue worthy of a prime-time soap opera. Loewy loathed the anti-European bias that the car guys cultivated, and Exner, who had yet to truly put his mark on an entire line of cars, felt Loewy was holding him back. Exner soon found himself reporting to Cole, the dominating chief engineer. At the time, executives Paul Hoffman and Harold Vance were deeply involved with wartime responsibilities in Washington, DC, which left Cole with much more time to make his presence felt in the factory and in the boardroom. Cole felt that Loewy was featherbedding his design hours, racking up needless overtime and superfluous billings. The Studebaker design contract gave the company a set number of hours from Loewy's firm. Any hours beyond that meant overtime. Cole also knew that company president Paul Hoffman had authorized Loewy to double his out-of-pocket expenses. Cole seethed as he watched the designers seemingly drag their feet on the postwar rollout and run up huge tabs at South Bend restaurants.

Virgil Exner Jr., who also became a car designer, in an interview decades after the Loewy-Exner confrontation, framed the argument from his father's perspective. "The trouble was promoted by Roy Cole because even my father had trouble getting Loewy out from New York to take a look and make any kind of commitment to the postwar car development. He would show up and make rambling non-decision decisions and take off and soak Studebaker with an expense account."

"My dad knew that he'd submitted his expense accounts and he tried to be very honest about the whole thing. Loewy would take a cut on them. Studebaker told my dad that Loewy was taking a cut on this," he continued.

Exner's frustrations with Loewy's scarcity and a festering resentment that Buehrig had an equal title in the design group eroded Exner's loyalty. Cole felt a bond with Exner, the hometown boy and enthusiastic car guy. Designer Holden "Bob" Koto recalls the intrigue: "For several months we were working individually on the package. Unbeknownst to us and to Loewy, Exner was working on yet another package in his basement at home. All our separate models in the studio were cast and we had shows for top management, and meanwhile Exner and Roy Cole had Budd [Co.] in Philadelphia build the full-size wooden mockup of Exner's basement design. The Budd mockup was shipped to South Bend. Management bought it—accepted it on the spot."

Exner's off-site design was a classic "skunkworks," a term popularized in the aircraft industry to describe research and development teams within an organization that have a great deal of autonomy (the term derives from the moonshine operation in the comic strip *Li'l Abner*). The backstory for Exner's skunkworks dates to 1944, when Cole asked Exner to work on the postwar model on his own. Exner would come into the studio every morning at about eleven o'clock, saying he had been consulting with Cole in the engineer's office, when he had actually been up late working on his Cole-commissioned car. Cole encouraged Exner to break with Loewy, which Exner would not do. But he would work on a competing model for the 1947 in his basement. Cole, who perhaps studied at the Niccolo Machiavelli School of Engineering, went so far as to give Exner the correct dimensions for the 1947 car while giving Buehrig false numbers. Later, Cole told Buehrig that his objective was not to undermine the Loewy gang, but to get rid of Raymond Loewy.

Cole's admission of giving the other designers the incorrect specifications was a last-minute effort to get Buehrig to defect to his camp. Buehrig refused and hurriedly reworked the model for the Loewy team for the presentation of the new design to the board of directors. The dimensions of the Buehrig model were still off, and the executives chose the Exner model. Loewy, interviewed years later, laid the blame for the intrigue at Exner's feet. "I soon realized that Exner's conception of advanced body styling clashed with the ideas of the other fellows on the team." Loewy, who always made the final design presentation for all clients, found out about the backroom intrigue at the presentation. Incensed, he fired Exner. According to Exner, Loewy said, "I will have nothing more to do with you. You are immediately fired." Loewy would remember later in an interview, "In my experience in fifty years, working in more than one hundred corporations, I have never seen such a case of despicable behavior." Cole was present at the meeting, and several people recall him saying seconds after the Exner firing, "You are immediately hired Mr. Exner, by the Studebaker Corporation."

Bob Bourke, with characteristic understatement, summarized the incident, "It was sort of an underhanded deal on the part of Roy Cole, because he was trying to get Loewy out of there." After the confrontation with Loewy, Exner was set up with an office on Studebaker's Proving Grounds, the off-site testing facility on the outskirts of South Bend. Two weeks after the confrontation, Loewy fired Buehrig, mainly because the older designer had failed to tell him about the poisonous atmosphere in South Bend. Loewy "was very temperamental and these other people were temperamental," Audrey Moore Hodges said of the tension between Exner, Buehrig, and Loewy. "It was just a clash of temperaments. I never saw anything. I just noticed that one day they weren't there."

After the bloodletting, Loewy made Bourke Studebaker's chief designer. Cole, never content to win just one battle, forced the Loewy group out of the engineering building and into a former gymnasium near the plant. Even later, Cole forced another move to a local dealership. "We had to wheel our full-size clay models through the streets of South Bend to show them to management," said designer Tucker Madawick.

Hodges said that the Loewy designs often were compromised by the powerful engineering department dominated by Cole. "We would be invited to a showing of one of our designs. We'd hardly recognize it. Every engineer featured himself as a designer, and they would take some of the most well-intentioned designs—I can just see Exner rubbing his head so many times over what engineering would do to a perfectly beautiful design."

Loewy realized that he had become a lightning rod for controversy at the South Bend plant and began to cede more "customer service" duties to Barney Barnhart, who ran the transportation division in the Loewy New York office. Barnhart would do more and more product presentations, although Loewy did not remove himself completely. He worked out a series of presentation signals to make product meetings go more smoothly. If Loewy were asked a question he couldn't answer, he would drop his chin, prompting Bourke to jump in to answer the question. If Loewy had to leave early or felt the room turn against him, he would have the home office phone in an emergency call. His presentation skills, well rehearsed ahead of the meeting, almost always closed the deal. In truth, Loewy's distance from South Bend was not a great loss. Bourke said Loewy's great gift as a car designer was as an editor. "He couldn't draw cars. He was just ridiculous and he knew it," Bourke said. After the design was created, it was up to the design impresario to move the design from sketchpad to manufacture, certainly a harder job than drawing a car.

THE NEW POSTWAR Studebaker model was the "great leap forward," balancing the length of the hood and rear deck, which inspired the long-standing joke that "Studebakers don't know whether they are coming or going." The car featured some interesting "firsts," including a curved rear window, a deemphasized radiator grille, and a design where the hood and rear deck were part of the same body line. Loewy, in the automotive press and in company advertising, received full credit for the 1947 design. "Studebaker made Loewy a household word," wrote one design historian. In reality, Loewy had made his national reputation well before the Studebaker designs, but it certainly legitimized his work in the eyes of many automotive executives. The 1947 design was ultimately a mix of ideas from Exner and Bourke, who would eventually inherit the mantle as Studebaker's most influential designers.

Studebaker executives acknowledged the groundbreaking design in the 1946 annual report: "The management's decision to introduce genuine postwar models as quickly as possible based on the conviction that the company stood to gain much from being the first to give its customers the advantage of advancements both in design and production methods accumulated during the war and which, in total, represented substantial progress. It is quite evident from the public's reaction to the 1947 Studebaker . . . that this decision will prove to be one of the most significant in the company's history."

Design historian Edson Armi points to the 1947 Studebaker as a critical turning point for American car manufacturers. "Loewy's studio fundamentally changed the direction of American cars from the wartime monocoque to the European lightweight, tightly skinned and minimally chromed." GM's Bill Mitchell, ever competitive and eager to use Loewy as his personal bête noir, claimed dominance in the chrome wars in an interview with Armi. "I wasn't for chrome either. Where we broke the ice was when we did the first Riviera. That set the standard with hardly any chrome." Unfortunately for Mitchell, the 1963 Riviera still sported more chrome than Studebaker's 1947 model, not to mention the 1953 Starliner. After the war, the Studebaker became a design benchmark for the Big Three. Loewy pushed executives in South Bend to expand the design envelope, but the executives found it difficult to ride waves ahead of its larger contemporaries. The management team consistently questioned marketing decisions and Loewy's design direction. "The independent, in order to succeed, must be courageous and progressive. The results may be somewhat of a shock, but it is far better than blindness," Loewy said.

The new design, praised by auto writers as "forward leaning," led the way to the company's best sales years. The net sales figures for that model year totaled $141 million (roughly $1.56 billion today). By 1950, Studebaker had 4 percent of the domestic car market. The 1950 model year was the high point for Studebaker sales, moving 268,229 cars out of showrooms.

Henry Dreyfuss, always honest in his assessments, described the 1947 design favorably. "Raymond Loewy's first postwar Studebaker was a car that reflected its honesty, and, as I see it, has led the whole industry several steps along the path toward intelligent design."

The fallout from the executive intrigue surrounding the 1947 design was extensive. Exner worked alone in the luxuriously appointed design studio, while Loewy and Bourke, who were contractually obligated to Studebaker, worked on design problems. Loewy and Bourke worked on the redesign for the 1950 model.

The 1950 models updated the 1947 look, sporting the "spinner nose." Loewy designer Vince Gardner, inspired by the other end of the P-38 fighter plane, urged Bob Bourke, now chief designer, to put a spinner nose flanked by forward-leaning fenders that aped the double-fuselage look of the warplane. The spinner grille was a direct steal from the conical propeller assembly of the P-38, P-39, the P-51, and others. The spinner nose was complemented by two chrome moldings set at nine o'clock and three o'clock, which quickly drove the airplane allusion into the mind of the consumer. Exner hated it, and Loewy was lukewarm about the aesthetics of the design.

Most design critics either love it or loathe it. Historian Thomas Bonsall says, "The bullet-nose Studebaker is almost unique among his production cars and being an example of the 'private' Loewy." The 1950 Studebaker line of all models set company sales records, moving 329,884 units, which was still only good for ninth place among all American carmakers. Loewy himself was ambivalent about the 1950 car: "The result was a bulbous, rather clumsy, fat automobile. Aesthetically, I never liked it much, but the customers did."

Loewy claimed he had thought about introducing this design theme as far back as 1942, but designer Bruno Sacco doubts that was the case. "Still there is one undeniable forerunner of the bullet-nose concept, his S-1 locomotive for the Pennsylvania Railroad," Sacco wrote. With the spinner grille debut in 1950, critics decried the excess chrome on a groundbreaking design that had largely been devoid of "schmaltz."

Although Loewy's signature graced all the Studebaker design sketches, it is clear he did little hands-on designing for Studebaker. In the era of the annual model update, design studios created sheaves of drawings not only of the model's look, but also of details such as hood ornaments, bumpers, and side mirrors. Drawings were used less to market a car to the public than to sell ideas to company executives. Designer Bob Andrews explains: "Sometimes people criticized [Loewy's]

The 1950 Studebaker Land Cruiser in a glowing publicity photo. The famous bullet-nose design emerged after a feud between Raymond Loewy and Virgil Exner over design credit for the 1947 Studebaker. Both Loewy and production executive Roy Cole expressed low opinions of the design, but the airplane-inspired look became the iconic symbol of the cars produced by Studebaker long after the company shut its doors. (From the Collection of the Studebaker National Museum, South Bend, Indiana)

Hollywood approach. I never did. They'd say 'Mr. Loewy, he signed your drawings.' That was understood when you went with him. His theory was to come into a board meeting of a plant company and to be more famous than the chairman of the board. He was much better known and had an established accreditation. That would be very difficult, because many people that come up through the managerial status—through finance—have a great distrust and dislike for creative people.'

Even with Loewy's implicit approval of ladling on the schmaltz, the spinner-nosed 1950 Studebaker did not spin off extra sales. While it was still in preproduction, Loewy's nemesis, Roy Cole, remarked, "Boy, if they ever come out with a silly thing like that we'll go bankrupt." Cole's disparagement didn't stop the company from marketing the new design. An excerpt from the company's advertising called the Land Cruiser "a melody in metal, a symphony in steel, beautifully streamlined, excitingly new." The spinner-nose car was celebrated in film as well—fans of the 1955 British comedy *The Ladykillers* will recognize the Land Cruiser as the getaway car used in the film's central robbery. The original 1947 design and its subsequent redesigns is today universally recognized as a breakthrough. A 1960 *Esquire* article by racecar driver John Cooper Fitch notes, "The 1947 Studebaker bore the clean imprint of Raymond Loewy's designing genius. Its lines were graceful, long and a little hungry-looking, compared to the bulging smugness of prewar cars. There are those who prefer Loewy's '53 model, but, relating each to its time, I must stay with the '47, for that was the first quivering step towards the future and an opening dividend on that promised world-of-tomorrow." Loewy quoted this line verbatim in *Industrial Design*.

The final story in the triumph of the postwar Studebaker was its inevitable "knock-off" by one of the Big Three. The 1947 Studebaker design was not on showroom floors more than a few days before the Big Three sought to mimic its innovative design. Ford, which was the most successful at emulating the lessons of the Champion, had an inside track. Shortly after the new model's debut, Bourke had been forced to lay off the underutilized Dick Caleal, a lower-level designer who had just started at Loewy and Associates. Bob Bourke recalled that Caleal had decided to stay in South Bend despite having been let go because his wife had a local job. Caleal, an Arab American, had spent time designing at GM, Packard, REO, and Hudson before being hired at Studebaker.

Ford Motor Company emerged from World War II ill prepared to meet the market. After years of tyrannical rule by Henry Ford, who was loath to improve a product he already thought was perfect, little research and development had been done. The design studio had been cut to bare bones during the war. Henry Ford II, known in Detroit as "Hank the Deuce," had finally wrested control of the company from a cadre of Henry Ford's cronies in 1945. Desperate to bring the company out of the past, he looked to save the company by hiring Ernest Breech from General Motors in 1946, followed by engineering executive Harold Youngren. Breech wanted the new model to be ready for the 1949 model year, meaning he had roughly twenty-four months to produce a new car. Breech did not respect Ford's designers, so he used the short production schedule as a means to bypass the in-house staff.

The compromises Studebaker failed to make in the 1947 Exner-designed car by pitting two design teams against each other were finally realized as a coherent design. Unfortunately, the breakthrough came at the Ford design studios. The in-house assignment went to designer Bob Gregorie, but Breech also contracted with independent designer George Walker to create another version. Walker in turn hired Caleal to execute the Ford design. Caleal, who was living in Mishawaka, Indiana, designed the winning Ford model, aided in secret by Studebaker designers Bob Bourke and Holden Koto. To say that the 1949 Ford was inspired by the postwar Studebaker is an understatement. The car was nearly a carbon copy of the Bourke team's Studebaker design. When Gregorie added a more rounded profile and changed the line of the rear and taillight designs, the 1949 Ford found its own personality, although its bones and structure were clearly inspired and copied from the groundbreaking South Bend design. Caleal's Ford design was referred to as "the car that saved an empire." The car generated more than one million sales and $177 million in profit.

Before the '47 Studebaker and the '49 Ford, all cars had bulging body shapes, particularly around the front and rear fenders. Walker claimed authorship of the Ford design by claiming, "We smoothed those lines out and began the movement toward integration of the fenders and the body." The duplication went well beyond aesthetics. According to historian Armi, both Ford and GM bought a Studebaker new model. Ford engineers disassembled it and weighed and tagged each part, giving Ford at least a virtual blueprint for the specifications for the 1949 model. The Caleal and Gregorie designs were both presented to Ford executives, and the Caleal design was chosen to go forward. The 1949 Ford used elements from both design teams, including a spinner grille designed into the chrome bumper and radiator, which Loewy would appropriate a year later. The battleship-heavy prow of the Ford even out-chromed Harley Earl's designs, although the proportions were more balanced than GM models (after the debut of the car, *Time* magazine called Walker "the Cellini of Chrome").

Caleal emerged as a hero and was made head of advanced styling for the Ford design team. Walker, who had an obvious vested interest in claiming credit for the design, remembers Caleal's contribution differently. "He did so little. He did clay modeling for us, that's what he did. But he was a blackboard man, when they took the line and he interpreted it onto a blackboard so they could make an engineering drawing. That's all he knew, you see? Dick is not a designer. He wasn't even a clay modeler." Frank Bianchi, a Ford designer for forty years, recalled, "Dick had plenty of help, Joe Oros and Elwood Engel had an awful lot to do with that car." Walker's disparaging remarks aside, all the Studebaker designers recall Caleal and his extensive contributions to the winning model. "[Bob] Koto worked on the sides and the roof, and he worked on the windshield, the window details. Did a little clay work myself and the front-end detail was primarily me," said Bourke. "Dick did a lot of the finishing work on the car, all the way around. I'd say he contributed as much of that thing as anybody, if not more." Caleal would later go on to work for Chrysler as studio director and ended his career as head of the Dodge Truck Studio.

In contrast to the effective-yet-modest marketing efforts made by Studebaker, Ford rolled out its new design with a $10 million campaign that began with a weeklong debut in New York City at the Waldorf-Astoria Hotel. The rollout crystallized the difference between an independent manufacturer and one of the Big Three automakers. Ford committed to a multimillion-dollar marketing scheme, while Studebaker management had to cut corners for its model debut. The new Ford was seen by more than 10 million consumers during its tour, resulting in Ford sales of more than a million cars in its first model year. The accelerated production schedule at Ford caused quite a few fit-and-finish quality problems, but the 1949 and subsequent models put the company in even competition with modern, forward-looking designs.

Exner left Studebaker after Cole's retirement in 1949 and went to Chrysler as the head of the company's Advance Styling Group. His reputation had already been made at Studebaker, but Exner's legacy was sealed at Chrysler. Rather than designing the Chrysler production cars, Exner created a series of show cars and prototypes. In collaboration with Carozzeria Ghia of Turin, Italy, Exner created automobiles that pushed the envelope of design. Chrysler needed that push toward more innovative design because company president K. T. Keller had long insisted that men should be able wear their hats while driving Chrysler products. This "old hat" philosophy produced cars that were boxy and unappealing. When he became director of styling in 1953, Exner went on to define the stylistic excess of the 1950s in his "Forward Look" line of cars. "We wanted in the Forward Look cars an appearance of fleetness, the eager poised for action look, which we feel is the natural and functional shape of the automobile."

Taking the forward look of his Loewy studio 1947 Studebakers and exaggerating the angles of the rear deck and front grille, Exner created an increasingly baroque series of cars that used prominent tail fins to separate Chrysler from the rest of the car companies. His signature look included flaring tail fins, ample glass, lean roof lines, minimally decorated sides and grilles, and restrained interiors. The 1955 Chrysler 300, with its chassis poised to take off like a sprinter in the starting blocks, was a seminal design model for the ever steeper tail fins to come. Exner also introduced several concept cars, including the 1952 D'Elegance and the 1956 Norseman. The Norseman was a product of a collaboration between Chrysler and the Italian designers Ghia. The car featured a retractable sunroof and a cantilevered roof extension shading a wraparound windshield. The slender supports used to form the roof offered the Norseman nearly all-around visibility. The final prototype was loaded onto an ocean liner for deliver to the Chrysler design studio. But the *Andrea Doria* collided with a cargo ship off the Massachusetts coast, and the Norseman sank to the bottom of the ocean. Exner, who had suffered a severe heart attack right before the sinking, was not told about the accident until he had more fully recovered.

The 1957 Imperial might have been Exner's design benchmark, featuring curved glass and rakish fins. It also debuted "gunsight" taillights (inspired by the sights used on .50 caliber machine guns) and a spare tire cover pressed into the trunk lid. It is considered one of Detroit's most memorable designs. The fins were nearly deleted in the concept stage, causing Exner to exclaim to his wife, "My God Mildred! They want to cut [the fins] because they can save 36 cents!" After 1957, Exner, the man who famously resented Raymond Loewy's taste for the spotlight, was featured throughout the company's advertising. Exner tried to dole out appropriate credit, but in the public eye, he was another "genius designer." Exner defined the market throughout the 1950s with a series of increasingly baroque finned models, but he was fired in 1962 when a series of shortened sedans, ordered by Chrysler executives to compete with GM and Ford designs, failed.

Once Roy Cole retired and Loewy consolidated his design team after the Exner debacle, Loewy felt more confident in maintaining the Studebaker account. "As he wrote in *Industrial Design*, "Thanks to Paul Hoffman, I was given the opportunity to design cars liberated from most of Detroit's atavistic style. No more inbred, incestuous designs; instead a fresh new approach for a century-old respectable firm was demanded. The body-styling division, which I formed at the plant and that bore my name became known in the profession for its talent, spirit and sense of mission." His main mission, as he saw it, was to reduce vehicle weight and to center the design team's ideas on the chrome-free lines of European sports cars.

9

The Starliner Coupe
Studebaker's Breakthrough Design

Describing a new red-and-cream Studebaker hardtop zipping down the Studebaker test track, *Time* magazine spared no adjective. Documenting every twist and turn, the writer found profound drama in every lap. "For six months the car was driven, in well-shrouded secrecy, until it had piled up more than 100,000 miles. Not until then did Studebaker Corp. engineers feel that they had worked all the bugs out of its 1953 car."

Described as the nation's most adventurous automotive design since the debut of the 1934 Chrysler Airflow, the magazine touted Studebaker as the first automaker to market a "semi-sports car." The 1953 Starliner would become the inspiration for the benchmark sports cars to come. More accurately, the coupe was the forerunner of the "personal luxury car," designed for suburban executives to drive to the office. The automotive market was changing from the old one-car-per-family dynamic, where Dad drove a large, conservative sedan that could take him to work and also haul groceries or take the family to church on Sunday. After the war, as incomes and families expanded, families often owned two cars, a larger "people hauler" for Mom, and a new type of car that the head of the family could drive but did not use for family-oriented activities. From the Starliner evolved the Thunderbird, the bigger Buick Riviera, the even larger Oldsmobile Toronado, and a host of other makes. Led by Studebaker chairman Harold S. Vance, the company dedicated $27 million to retool the South Bend plant and scheduled an unheard-of

40 percent of its manufacturing production to the racy new design. The model was available in a five-passenger hardtop ('50s auto talk for a car with a two-tone roofline) and a coupe, retailing from $1,800 to $2,300.

Time credited the new sports car's look to "the designing skill of Raymond Loewy, a sports car owner himself." In reality, Loewy was more of a sports car dreamer. He often bought European sports cars and had mechanics and body men customize them to his specifications. Some of his creations were outlandishly over the top, but when clients had the chance to translate these designs to a popular production car, the dialed-down version often sold well. Loewy's ability to create one-of-a-kind prototypes out of existing models made him the perfect designer to create an American sports car with a European sensibility. It also did not hurt that he had traveled throughout Europe—unlike some other American car designers, whose Big Three–centric worldview made American car design prior to 1953 so insular. *Time*'s article, which also featured color photos of new sports coupes from Nash-Healy, Jaguar, MG, and Porsche, identified a new trend for Americans looking to buy sporty, European-influenced sports cars. Of course, by the time the magazine identified the trend, it had been in motion for some time. *Time* failed to anticipate the nation's brewing love affair with automotive horsepower and performance. "Despite all the new cars, no American automaker thinks there is a big market for a true sports car in the U.S.," wrote the article's anonymous author. Vance agreed, but his idea was to create a stylish sports car for the family. The question he should have been asking was: Would enough families react to such a stark change in what buyers regarded as a family car? High-end sports cars would always remain a niche market in the United States despite *Time*'s prediction, although the "family sports car" concept pioneered by Studebaker would come to fruition in the 1960s.

Loewy's autobiography had already been published by the time of the Starliner's debut, so he did not have the opportunity to embellish his design legend at the expense of others. With the car industry being so closely covered by the automotive and business press, Loewy knew he could not create "print the legend" stories of cocktail party insights or meetings of the mind between Loewy and CEOs. The 1953 Studebaker Starliner started its life in 1951 as a Bob Bourke–designed show car. Bourke, Loewy and Associates' most consistent auto designer, took great pains to keep the car's dimensions within existing production standards.

The 1953 Studebaker Starliner, designed by Loewy's senior automotive designer Robert Bourke, brought Loewy's dictums of lighter, lower, and less bulky to the US car market. The coupe was incredibly popular with buyers seeking a "personal luxury car," but Studebaker could not keep up with the demand. (Raymond Loewy™/® by CMG Worldwide, Inc. / www.RaymondLoewy.com)

"I always wanted the chance to show what could be done if we didn't have too many restrictions," said Bourke. "During this period I worked seven days a week plus three or four nights a week . . . I recall once working for three days with a total of six hours sleep."

Another reason Loewy did not engage in self-aggrandizement with the latter-day Studebaker designs is that he was friendlier with top management. In 1953, Vance knew instinctively that Studebaker could not match the marketing power of the Big Three. He admitted as much to *Time*, saying the car had to sell itself. Loewy underlined this thinking with a sophisticated quote to the magazine, "We knew it would [sell] if it would be fresh and gay and young-looking—what the French call witty." What the American car buyers called it was a perfect blend of comfort, economy, and looks.

Car designer Bruno Sacco, who helped create the look of Mercedes-Benz cars from 1958 to 1999, noted that the 1953 model was Loewy's triumphant design for Studebaker—the process was remarkably free of the infighting, struggles, and business setbacks that plagued many of his other designs for the company. "The 1953 Studebakers could not have been produced without a total consensus between engineers, technicians, and designers: that these models—particularly the Starliner—were developed, built and marketed shows that all concerned were of one mind and believed in what they were doing. This was a triumph for Loewy: the entire concept of the automobile was based on the design," Sacco wrote. The claim for the Starliner as a new paradigm for American cars is not a hard case to make. In the early 1950s, General Motors and Ford were still producing massive rococo-chrome monsters, and the sleeker designs of Chrysler and Ford were still a few years away. The Starliner was more a progeny of the European designs from Ghia and Pininfarina. Sacco compares the Starliner favorably to a landmark design that debuted in 1954—the Mercedes 300 SL "gullwing" coupe.

Whatever the inspiration, the wants and desires of the postwar car-buying public were changing, often in ways car companies had difficulty predicting. The cars of the 1950s brought gadgetry and gewgaws into the automotive vernacular. Automatic transmissions became standard equipment. Push-button controls ran everything from windshield wipers to the drivetrain. Most of the advertising and marketing materials aimed these options at women. The automotive industry found that women often heavily influenced the decision to buy a car, according to a 1947 article in *General Motors Engineering Journal.* Ford had six female designers on staff by 1947. GM's design studio hired its first female design employee in 1948 and had nine on staff by 1958.

Loewy and his Studebaker design staff anticipated how cars would look for the next three decades more presciently than Harley Earl. While Studebaker was remaking the American coupe, Earl was creating perhaps the most outré concept car ever designed. The 1951 Buick LeSabre, inspired by the F-86 Sabrejet, featured a central intake in the front of the car, offset by two rubber-tipped "Dagmars" beneath the gaping maw. The turbine motif was continued through the back of the car, which was framed by two immense, bulbous fins. The single stylish feature of the LeSabre was its sleek, panoramic wraparound windshield, which would become a mainstay of GM's design studios for the next twenty years.

THE CARS COMING OFF THE assembly lines needed to appeal to consumers who had an ever-widening spectrum of choices and epic lists of errands to run, all of which required mobility. The 1950s inspired a flight out of the metropolitan city limits to the suburbs beyond. The migratory urban families required housing, and the construction industry built more than one million new homes per year, accompanied by roads, highways, and shopping centers. In the postwar years, General Motors passed Procter & Gamble as the nation's leading advertiser. In 1956, the leading individual advertising budgets were Chevrolet ($30.4 million, or $263 million today) and Ford ($25 million, or $227 million today), followed in order by Buick, Dodge, Plymouth, Mercury, Chrysler, Pontiac, and Oldsmobile. Only one nonautomotive company, Coca-Cola, was in the top ten. In 1945, 31 million cars were registered nationally. By 1960, that number jumped to more than 73 million. Gross advertising, driven by the car industry, whose products were differentiated primarily by appearance, rose by 75 percent from 1950 to 1960.

At roughly the same time, European sports cars were making inroads with well-off car buyers. There, most sports cars, built to corner and hug the twisty lower-quality roads crisscrossing Europe, were as comfortable as a Conestoga wagon and often provided a rougher ride. American consumers preferred a smoother automotive experience, best described as riding on pillows. American manufacturers were looking for a sporty car model that would both offer comfort and stand up to the heavy use and longer distances that American families typically traveled. Studebaker was the first to see a market for a sports car that could accommodate a family (or at least part of a family). Vance, a conservative businessman who could deliver the most bloodless of quotes, said dryly, "Originally we thought that our sports car would appeal only to younger people. Now we're finding to our surprise that it's appealing to all classes of people."

The genesis of the Starliner coupe is nearly as convoluted as the postwar 1947 model. The time line of the design has been well documented in memoirs and oral histories, leaving little room for Loewy legend polishing. Still cautious after the intrigue and recriminations of the Exner-Cole-Loewy triangle, Loewy reacted to Vance's assignment for the new coupe by delegating the design work to Bob Bourke and leaving for Europe.

Bourke started working on production designs in 1951. The design crew tackled the new assignment in addition to the design updates on existing Studebaker models. Vance took an uncommon interest in the

Raymond Loewy takes a shaper in hand to work on a clay model with George Matthews (*center*), and Robert Bourke. (From the Collection of the Studebaker National Museum, South Bend, Indiana)

process. He now dropped by the design shop often, when previously he would come by only when Loewy was in South Bend. Vance periodically cautioned the designers to watch their budget. Loewy returned from Europe and chose Bourke's design to present to Vance. A few weeks later, as a "show of force," Loewy, Bill Snaith, and Barney Barnhart presented the design the staff had chosen, and the Studebaker board authorized production. Bourke was given permission to start production on the new model design.

Bourke, Studebaker's most talented designer, with the longest career creating designs for the company, was born in Chicago, living in the same house until he moved to South Bend to begin work at Studebaker. As a youth, Bourke was fascinated by airplanes and built countless models, an interest he shared with his future boss. He attended the Chicago Art Institute and paid his way through school by mopping floors and cleaning bathrooms at the school. In the summer, he would work in his father's architectural office. Upon graduation, he worked for Sears at its Chicago headquarters, where he was mentored by Jack Morgan, a former GM designer. Shortly after Bourke's arrival, another designer, Clare Hodgman, was hired on at the office. Hodgman, who also worked at GM, had been hired by Sears to continue the designs he had done for Loewy on Coldspot refrigerators. Hodgman had done some early work on Loewy's Studebaker account and recommended Bourke to Virgil Exner. He joined Loewy and Associates in 1941.

Bourke's Starliner design was clearly inspired by the low-slung, minimally chromed sports cars of Europe, but the look remained uniquely American. Bob Andrews, another Studebaker designer responsible for the Avanti, explained how Studebaker retained a European influence in an interview: "The reason it was exciting is that Mr. Loewy had the only international design department in the United States at that time. We had Count Albert Goetz, John Cuccio. We had them from England. We had about every country [represented] . . . The thing that excited me was that it was the first time that I had an opportunity to compete directly with European designers and that was a hallmark. Each designer was given a quarter-size model and a modeler and we did our version of this new experimental car."

"I had felt for the last forty years that the American automobile was too bulky and heavy, necessitating large power plants [engines]," Loewy wrote in 1979 in *Industrial Design*. "I lectured about it, wrote articles about it, said it on the radio, and achieved only one thing: Detroit's resentment and hostility." Loewy had to wait two decades more to be vindicated by his downsizing campaign when the oil crisis of 1974 began a wholesale market correction to high-gas-mileage cars. "Detroit simply had no use for my ideas as long as the public wanted large cars with all their accessories, which Detroit could sell, inducing higher price tags," he continued. "And businessmen rarely express appreciation for ideas that may lower their income. In point of fact, they wouldn't have; while Detroit was building gas guzzlers, the Germans and Japanese, with their more modest cars, ran away with the world market!"

Well before Japanese automakers came into the American market, independent auto companies were struggling to make a dent of their own in the domestic market, but most were trying to beat the Big Three by building big cars. The Starliner was the best chance for Studebaker to make a significant splash in sales. Studebaker engineer Eugene Hardig is credited with the 1953 car's low profile, with a height of just 56.5 inches.

The 120-inch wheelbase was taken from the company's Land Cruiser model. When the coupes went into production, touted as having "the new European look," consumer reaction caught the company unaware, or at least poorly prepared. Production had anticipated that sedan sales would be double the demand for coupes. Over the next few months, fueled by the *Time* cover story and the public's avid acceptance of the design, demand for coupes was four times that of the company's sedans. The customer preference for coupes was understandable. First, the market was leaning toward the personal car. Second, Bourke's sleek coupe design applied to a longer, larger sedan was less than compelling. The company decided not to commission a separate design for the sedans, so the Loewy staff quickly adapted the coupe design for the larger machine. The sedans were ungainly, and the style that appeared sleek in a coupe seemed dumpy when applied to a four-door car. Sedan production for the Starliner fell to just 20 percent toward the end of the model year. The lack of sales for the new line of sedans was a crippling blow to the company's finances because their production line was set up to build more sedans than coupes.

The 1955 Studebaker coupes roll off the final assembly line at the South Bend, Indiana, plant. Although the Starliner coupes were highly praised by critics and sought after by consumers, Studebaker failed to capitalize on their popularity because its production lines were set up to build sedans, and the company could not retool fast enough. (From the Collection of the Studebaker National Museum, South Bend, Indiana)

The Studebaker sales force had to turn away buyers while the plants refitted production to the Starliner coupe. In addition, the company did not use common parts to lessen expenses and tooling costs. The company was further weakened by a price war started by Henry Ford II's decision to begin a production push, flooding the market with steeply discounted cars. Studebaker was further affected by the price and production war with Chevrolet engendered by Ford's retail move. Both companies were selling new cars for $50 or $100 above dealer cost. The price war between Ford and GM, which left the two large automakers unscathed, instead sounded the death knell for many of the country's remaining independent carmakers. Studebaker sales for the 1953 Starliner were further hurt by a major quality problem. Almost every company model experienced extensive rust issues owing to a substandard asphalt undercoating. The company eventually switched to a saltwater-proof compound, but for years afterward, the fender wells were still treated with the asphalt treatment.

Despite such quality-control problems, the public response to the new design was overwhelmingly positive. Loewy's auto design philosophy—always at odds with the dealers—was encapsulated in a 1940 speech he made to the Society of Automotive Engineers.

> We practical designers deplore just as much as the engineers the excessive bulk and weight of the modern passenger automobile. But instead of blaming the designer, why not single out our friends of the sales department. I know it sounds like first-class buck passing, but how many times have we heard the dear boys say to us "What the public wants is a big package," and we know that a big package means a great deal of weight and a lot of money?
>
> Weight is the enemy. The average automobile weighs thirty-five hundred pounds. Thirty-five hundred pounds of materials to transport one or two people just does not make sense. Statistics show that ninety-two percent of the cars on the highways travel with empty rear seats. The weight trend in the past years, I believe, has been decidedly retrogressive. This must change.

Loewy constantly pushed Studebaker to concentrate on smaller models. Studebaker tapped into the zeitgeist and introduced a smaller car at what seemed like the perfect moment, a strategy that allowed the company to leap back up the sales charts. Although Studebaker sales were always erratic from year to year, the debut of the new 1953 models allowed the company to escape a downturn for a few more years, primarily because it still had an active network of dealers throughout the country. "Unfortunately Studebaker was not big enough a car company to set automotive styles," wrote design historian Thomas Hine. As Loewy won kudos from the automotive press, Harley Earl could introduce fenders with Dagmars and sell millions. Hines again: "It was perhaps quixotic to make a lean pure automobile in the same year GM was developing breasts."

Mindful of the limits of automotive executive's imaginations, Loewy always showed the firm's sedate designs first, and when he sensed acceptance or approval he would unveil the staff's more forward designs. The presentation of the Starliner's hardtop convertible look took place at just such a meeting. Although the 1949 Buick hardtop first featured the faux "convertible look" production car, Loewy's firm applied it to a sporty coupe. Executives had wanted an actual convertible, but Loewy's market research revealed that few customers wanted to go to the bother of pulling the top down, even if they liked the look of convertibles. He showed the executives some conservative convertible designs and then revealed the Starliner hardtop. Loewy closed the sale. "His talent was selling the design to top management, a talent most designers lack," recalled Bob Bourke.

Bourke stressed that Loewy's talents were not attuned to on-the-board car design. In fact, Bourke said that Loewy rarely put pencil to paper, but when he did, he could gain the respect of his employees. "He would sketch to some degree with a soft pencil in a heavy-line style. The only hand line drawing I ever saw him make was one day when I saw him make a beautiful script. It had a beautiful flare and a beautiful swing and was very well executed."

Mechanically, the one-time quality of Studebaker models had been allowed to slip despite the styling advances of the Loewy crew, and sales executives failed to foresee the buying public's demand for more powerful cars. Still, other designers also recognized the Starliner's excellence. Henry Dreyfuss sent a note to Loewy that said, "Raymond—you deserve appreciation from all designers—for letting us hold up our heads in the world automotive market." The widespread acceptance for the Starliner played a part in prompting Ford to produce the Thunderbird in 1954 and convinced GM to approve Harley Earl's 1953 Corvette concept car for full production the following year.

The release of the Thunderbird offered an excellent example of a large carmaker reacting to the successful design of a competitor. Ford executive Lewis Crusoe saw the popularity of the Starliner and the Corvette concept car and ordered a two-seater "answer model." Hewing to a Ford directive that the car should have a straight line from headlight to taillight (to differ from the curvy looks of the Corvette and the Starliner), company designers bypassed market analysis, studio critiques with executives, and design conferences. The Thunderbird, plain but powerful, outsold the Corvette (probably thanks to its powerful engine) 24 to 1 at the height of its popularity, only to lose its market share when Ford redesigned it as a four-seater in 1958. As for GM, Earl had noticed the Jaguars, MGs, and Alfa Romeos dotting driveways around Detroit and other cities. Earl's first Corvette prototypes, code-named "Opel,"

An advertising "beauty shot" of the 1955 Speedster, perhaps Studebaker's most influential design. It represents a complete manifestation of Loewy's vision. Its low-weight, low-profile, high-mileage attributes made it a forerunner of the personal luxury cars that emerged when American families began to buy two cars—a people mover for mom, and a personal car for dad. (From the Collection of the Studebaker National Museum, South Bend, Indiana)

for GM's European subsidiary, created visually stunning cars that were relatively underpowered because engineers used standard Chevrolet components, including a six-cylinder truck engine. The same young men whose tastes were turned to performance cars in the 1950s were underwhelmed by the Corvette's power, as well as by Chevy's insistence on using a two-speed automatic transmission and drum brakes that required nearly superhuman effort to stop the car. The car sold poorly, and, according to GM histories, there was some consideration given to scrapping the model.

GM assigned a Russian-born engineer, Zora Arkus-Duntov, to overhaul the car. The Corvette was the first production car with a fiberglass body, which made it lighter and gave the company a shorter production time. In 1955, Chevrolet introduced its first V-8 engine, and Arkus-Duntov redesigned it to fit into the sports car. He replaced the automatic transmission with a three-speed manual transmission, and the Corvette was off to the races.

In addition to Loewy's smaller machines, the work of Exner, the one-time Loewy employee who smoothly incorporated larger tail fins into the overall design of the 1957 Chrysler line, jarred the design shops at General Motors. Chrysler's sales had jumped after the introduction of the Forward Look line of cars, and GM executives thought it would lose significant market share unless it went fin to fin with Chrysler. Earl, in the last years of his long tenure as design chief, and Bill Mitchell, in his first years of oversight, decided to outflash the competition. The fin war reached its pinnacle in 1959, with Cadillac cars sporting the largest fins used on a GM production vehicle. Earl combined two passions in the design by including a cylindrical pod at the fin's midheight, mimicking the jet engines used in rocketry and jet fighters. Earl had a hand in the design, which debuted in 1958, the year of his retirement. Mitchell officially took over the design studio the same year.

AFTER THE DEBUT of the 1953 designs, Loewy was at the height of his popularity and influence as a car designer. Writers sought him out for opinions. Throughout the 1950s and early 1960s, Loewy was extremely competitive with his counterparts in Detroit and never missed an opportunity to wittily jab Big Three designs. He called Detroit designs "cars filled with spinach and schmaltz" and "mastodons." After being forced out by Studebaker, Loewy made a speech in 1955 to the Society of Automotive Engineers in which he described the typical American car as "a jukebox on wheels," and asked, "is it responsible to camouflage one of America's most remarkable machines as a piece of gaudy merchandise?" Henry Dreyfuss, who had little experience at car design, was even more disparaging of the Harley Earl school. "Detroit's reliance on the stylist is based on the perverse notion that the way to care for a sick man is to call in a tailor. Don't fret about his organic ills, just buy him a new suit . . . he'll look better. For all the emphasis on styling our cars are vulgar and monotonous. They're like the legendary brassy blonde: She's dazzling for the first five minutes, but then you're embarrassed to be seen with her."

While Studebaker owners weren't embarrassed to own the company's wares, the sluggish sales of the overall Studebaker line inspired the company to search in earnest for a merger partner to stimulate the company's declining stock price. The 1950s devastated Studebaker, but all carmakers were losing bits of market share to newly arrived imports such as Volkswagen, Mercedes, and others. Most automotive journalists credit the inroads made by imports to economical performance and inexpensive price tags. Perhaps more significantly, the often oddly styled imports offered a contrast to the sameness of Detroit metal. Smaller cars were easier to drive, compared to the average American land barges. Studebaker was aware of some of these factors but was powerless to address most of them. In 1954, the company merged with Packard, a former luxury brand whose models had failed to keep up with the marketplace. The company offered board positions to Harold Vance and Paul Hoffman, the most visionary of Studebaker's many leaders, essentially putting them out to pasture. Packard's CEO James Nance, who was brought in to revitalize the company, took over. The merger was friendly, but to appease stockholders, Packard took the reins. The Packard takeover was a universal cause for resentment at Studebaker, where employees referred to the years 1954 to 1956 as the "Packard Operation." Nance was purely a financial manager who had led GM's Hotpoint appliance division, and he was appalled that Studebaker's labor costs were twice the industry norm.

The Packard merger did not immediately turn things around, so Nance decided to compete with the Big Three by introducing a full line of cars ranging from economy to luxury sedans. Nance originally sought to accomplish a full-line strategy by merging with American Motors.

Instead, he decided to radically reduce costs by cutting Studebaker executive salaries and taking hard stances in labor talks. Coupled with this risky cost slashing, Nance had to compete for buyers at a time when car buyers had more purchasing options than ever before and were beginning to explore several imported brands. Suddenly there weren't just three giants and a handful of independent carmakers. Whatever Nance's strategic plan was, his numbers-only approach to automotive management was doomed from the start. L. David Ash, a celebrated designer for Ford, had this to say about Nance's managerial gifts: "Nance was one of the worst automobile executives I've ever come in contact with, for no other reason than he didn't know anything about cars. He couldn't even drive a car."

Nance felt that Studebaker did not have a broad enough range of models and that this lack of diversity made it difficult to ramp up sales and keep valued employees. Annual Studebaker salaries were $10,000 to $15,000 lower than those at GM and Ford. Talented engineers such as John DeLorean, who started his career at Packard and Studebaker, left as soon as he accumulated an adequate portfolio. His work on GM's GTO and other "muscle cars" revolutionized the industry, at least for a short time. Nance, an apostle of systems, found that internally many workers and executives were resistant to change. Most of his hires came in the finance department at a time when the company needed "car guys."

Nance's remaking of Studebaker using the GM system extended to Raymond Loewy's studio. When Loewy convened meetings to discuss the 1956 models, he came into direct conflict with Nance. Loewy had intended to introduce new features such as aviation dashboards, rack-and-pinion steering, and fuel injection. Nance vetoed it all on budget grounds. Nance accused Loewy of misleading the company by focusing on European styling, when in fact Nance hated the Euro-centric style Loewy preferred. Nance decided to ape GM and Chrysler by adding more chrome and larger grilles. Nance couldn't break Loewy's $1 million contract, but he cut Loewy's responsibility to minimal research and development—primarily going to car shows. "Our billings were about [$1 million] a year," recalled Bourke. "That included a lot of woodworkers and dry modelers, too. Loewy marked up everything. He marked up salaries 100 percent. I want to be very clear on this. It was entirely customary to mark up that much. But at the time, it was the largest industrial account and it was very expensive for Studebaker."

Bourke and Loewy's last assignment was the styling of the Hawk series. "It took Nance something like a year and a half to go through $4 million and at the end of that he had a couple warmed-over versions of nothing," Bourke commented years later. In 1956, the Loewy studio in South Bend was closed.

In a personal letter apparently written to either Paul Hoffman or Harold Churchill—the salutation reads "Dear Paul (or Harold)"—Loewy made an effective and eerily accurate prediction of Studebaker's future as he urged the executives to reconsider his ouster. "Nearly seven years ago, when the company could well afford the tooling cost, I felt the company's future lay in a field somewhat different than the 'Detroit Look,' already well covered by the Big Three." He went on to say that the company had squandered the popularity of the 1953 Starliner.

It is a tragic reality that this series of bad breaks practically extinguished the enthusiasm the new model had generated. In the period that followed, many explanations were offered for the failure to reclaim sales volume and among these reasons were many relating to design and styling. The atmosphere was one to encourage this type

Raymond Loewy and Bob Bourke look at a clay modeler working on Studebaker's 1955 sports car. The car, which features nascent tail fins, was never put into production. (From the Collection of the Studebaker National Museum, South Bend, Indiana)

of thinking, however contrary it was to the conclusions of those who had observed the situation objectively. It is therefore alarming to see the present policies of the Company switch so far in the opposite direction without apparently giving proper weight to the things that had been learned by hard experience.

He then staked his claim as the company's design savior, making an effective case: "More than ever I feel that Studebaker can have brilliant success by being the pioneer in types of motor vehicles that set design trends . . . Overdressed, square cars have reached their Waterloo." The final sentences could stand in for Studebaker's epitaph. "Again, survival for the independent demands that he find his own niche. HE MUST NOT COPY COMPETITION. NOW IS THE TIME TO STRIKE OUT WITH THE NEW CONCEPTS IN MOTORING. —THE PUBLIC IS READY!"

Loewy let a bit of bile loose after his ouster in an article in the *Atlantic*, where he wrote, "I am told that cab drivers have the highest rate of duodenal ulcers. I'll bet a chrome-plated carrot that automotive stylists have them beat. Every really creative and imaginative stylist and many engineers I know seem to be frustrated in their work today. The near-shattering pressure of their repressions is relieved in constant doodling—blue-sky dreaming. They rush home and make scale models in the attic. Or they long for the weekend to go road-racing in the old Isotta-Franschini or the souped-up pre-war Ford."

In 1956, Bourke transformed the successful coupe into the Hawk series. Directed by executives enamored of a schmaltz-laden line of Buicks to introduce more chrome, Bourke centered his design on a huge chrome grille that disturbed the once-unbroken lines of the Starliner coupe. The same year, a group of Porsche engineers offered a prototype, the Z-87, to Studebaker for a possible distribution deal. An assessment group led by John DeLorean vetoed the deal. In the end, Nance brought in another designer, Duncan McRae, a Ford designer. McCrae cleaned house of the rest of Loewy's group shortly thereafter.

At the same time, Nance was struggling to save Studebaker-Packard from liquidation. An influx of cash could be had with a possible distribution agreement with Daimler-Benz, the parent company for Mercedes automobiles. As Studebaker-Packard failed to make any of its sales goals, the company teetered on the brink of failure. Paul Hoffman appealed to President Dwight Eisenhower to save the company. Treasury Secretary

George Humphrey was called in to broker a deal. The complicated solution was to bring in the Curtiss-Wright Company to provide management help and to receive stock options that could afford Curtiss-Wright, if exercised, the chance to take over the car company. All Studebaker-Packard's defense contracts were assumed by Curtiss-Wright, and Daimler-Benz contributed $10 million to have Studebaker dealerships distribute the Mercedes brand throughout the United States. The distribution deal temporarily stemmed the exodus of dealers seeking more lucrative franchises. The result was far from smooth sailing, however. Studebaker mechanics could not work on Mercedes models, and the Daimler AG manufacturing engineers refused to add air conditioning, power steering, or automatic transmissions on any of the models destined for the US market.

Ultimately, Nance could not pull off the merger and resigned in favor of Harold Churchill, a Studebaker stalwart who had engineered the 1939 Champion. Although Churchill ran Studebaker, most decisions went to Curtiss-Wright CEO Roy Hurley. Hurley was a cost cutter, and the company tried to produce several cars on the cheap. Churchill also produced a bare-bones version of the Champion, called the Scotsman, to compete with Chevrolet and Ford in the low-price category. Yet instead of competing with the Big Three, the Scotsman was leaching sales from more expensive Studebaker models. Sales continued to drop in 1957 and 1958. The company expected Mercedes to sell briskly, but Daimler-Benz executives refused to raise their quota for American sales.

As Studebaker's financial officers struggled to make cars while simultaneously seeking other companies to buy in order to diversify its assets, the company's "car guys" were committed to introducing a new line of compact cars. In a rare case of predicting the market, Studebaker manufacturing executives decided that American drivers wanted smaller cars. Uncharacteristically, they were right, and national sales figures from 1956 to 1959 would bear them out. By 1955, Rambler (American Motors) sales were 80,000 for the small family cars. In 1956, the company sold 186,000 cars. Volkswagen began to make inroads into the US market, accounting for about half the import market in the late 1950s with sales around 28,000. By 1960, Volkswagen sales were 160,000.

Harold Churchill, promoted to president in 1958, was a lifelong employee of the company, working his way up through the engineering and manufacturing divisions. In other words, he was a car guy.

Churchill authorized the introduction of the Lark, a small car designed for women, who they hoped would buy it as a second car. The Lark, which was not a Loewy-styled car, was created from components of existing models. It was styled by Studebaker designers Duncan McRae, Virgil Exner Jr., and Bill Bonner. The car was introduced in 1959, just in time to catch the wave of interest in smaller cars such as Volkswagen, Volvo, and Austin. At the time, Studebaker joined Rambler as competitors in the small car market.

The market responded, with both makers sharing about 8 percent of the market and both showing major profits. Studebaker recorded $23 million in profit as buyers bought more cars, particularly the Lark and the Scotsman. Although sales jumped in 1959, it was short-lived. Executives at the Big Three could read the trends moving toward smaller cars, but swiftly switching production lines was not as easy an operation for the larger companies, which preferred making larger, more profitable models. Within a few model years, they had caught up. The gains made by Studebaker at the end of the decade were negated in 1960 when Chrysler introduced the Plymouth Valiant and Dodge Dart, GM brought out the Chevrolet Corvair, and Ford introduced the Falcon. Profits at Studebaker were slashed to $3.3 million in 1960. In the lean early years of the 1960s, Volkswagen had three hundred dealerships and sold more than 150,000 cars. Studebaker had two thousand dealerships and sold fewer than 100,000 units.

For imported cars to survive their first years in the US market, the most important element for success centered on creating accessible dealerships. Chances of financial success increased exponentially if the carmaker could establish enough dealers. For the Mercedes-Benz models, an expensive luxury brand that had not yet been discovered as an alternative to Detroit luxury cars, gaining two thousand dealerships across the country ensured its success. For the Studebaker dealers who survived the rollercoaster sales of the company's last years, the Mercedes agreement would turn the dealership owners who survived into millionaires in the decades to come.

In 1958, the company had discontinued the Packard marque and by the next year offered only the Lark and the Hawk (the update of Loewy's original Starliner). The smaller Lark had 44,000 sales in its first year and reached over 130,000 in 1959; it matched those figures the following year. By 1962, the company did what all car companies do when sales fall. They enlarged the car and offered more options and bigger engines. The Lark lasted until 1964.

The executives at Studebaker were buoyed by the Lark's sales, but held no illusions about joining the pantheon of big automakers. Instead, they were focused on making the company more attractive for a takeover or buyout. The Studebaker board, which had divided itself into camps that were pro-automotive and pro-financial, barely supported Churchill. Several board members were angling to become chief executive officer, which left the executive suites in South Bend in a constant state of instability. Churchill decided to bring in another freelance designer after the Loewy team finished its commitments. The designer chosen to replace Raymond Loewy on the Studebaker account was Brooks Stevens, a Wisconsin-born designer who was at the forefront of second-generation American industrial designers. He came to the account fresh off the acclaim for another automotive design: the Oscar Mayer Wienermobile. (He was often quoted as saying, "There is nothing more aerodynamic than a wiener.") But Stevens's experience with automotive design had a much deeper legacy. He was the designing eye behind the conversion of the Willys-Overland Jeep into the postwar consumer market. He also consulted on the design of the 1951 Kaiser-Frazer, an influential entry into the postwar car canon of streamlined sedans that was one of the first cars to feature a padded dashboard.

Ironically, Brooks Stevens was perhaps the only industrial designer to mirror some of Loewy's principles of image-making. Born into privilege (his father was an executive with Cutler Hammer, a circuit-breaker manufacturer in Milwaukee, Wisconsin), he studied architecture at Cornell University before entering the design field by creating "a spiffy-looking switch box" for his father's company. He had to fight for clients in the early days, recalling in a *New York Times* interview, "I had to fight my way in to talk to anybody in the '30s. I had to not only justify myself, but justify my profession." By 1940, Stevens, who had opened his own firm in Milwaukee, had five staff and seventy clients. Like Loewy, he had little patience for design as art. He was much more interested in delivering what the client wanted. He often derided the trend to think of designers as great artists, saying, "I decided at that point that an individual crusade for pride of art in the industrial product would be a thankless one as compared to the joyful ring of the cash register for my client."

According to Arthur J. Pulos, an American design historian, both Stevens and Loewy were showmen. Loewy presented the public with the archetype of the urbane, continental sophisticate, an image that often cost him jobs when dealing with manufacturers unused to working with a European émigré. Stevens also knew the value of a memorable first impression. He always arrived at meetings in distinctive cars, either new European imports or customized collectible cars. Late in his career, he created his own car, the Excalibur, an update and reinterpretation of a 1920s Mercedes-Benz touring coupe that left "more than a few parking attendants in awe as he drove by."

His clothing, like Loewy's impeccable dress, was intended to deliver the message "creator at work." At parties, Stevens would often entertain clients decked out in lemon-yellow slacks, a salmon sport coat, and a ruffled shirt. His home, a modernist dream built primarily of concrete in the International Style, intentionally separated him from his neighbors' French provincial and Tudor homes in the Milwaukee suburb of Fox Point. Some wags called the home "the only Greyhound Bus terminal in Fox Point." Stevens realized early on that he had to stand out from the crowd if he wanted to compete for business with designers based in New York. At the same time, he felt he had to have the same values as many of his clients, a belief that extended to his politics. He was a conservative, pro-business person who exuded the same beliefs as most of the Kiwanians and Rotarians of the 1940s through the 1960s.

Stevens also mirrored Loewy's belief that design is not art, but rather the art of presenting a product to increase its sales. Raymond Loewy displayed the slogan "It's good design if it sells" on the postage machine in the New York office. Stevens told the *Milwaukee Journal* about an account with the Evinrude Company, "We were out to sell boats to coal miners and lathe operators, not to connoisseurs in blue yachting coats." Stevens also went after accounts that were similar to Loewy's. He designed the streamlined *Olympian Hiawatha* train for the Milwaukee Road railroad.

Stevens's redesigns of the Willys-Overland Jeep are the spiritual inheritors of the modern-day sports utility vehicle and station wagon. Asked to create a new civilian model of the military Jeep, Stevens produced a sculpted model and design that bore little resemblance to the World War II workhorse. But this first pass at a redesign was tossed aside when the corporate president was ousted in favor of the more conservative

"Cast Iron Charlie" Sorenson. Sorenson had purchased a metal stamping plant that originally produced metal housings for washing machines. The limitations in width of metal meant that the auto body designers could not have any swooping, bulky curves. Stevens's solution was to create the world's first station wagon using the squared-off sheet metal that Sorenson's stamping plant could produce. The all-steel construction and two-tone pattern mimicked the "woody" look of prewar cars. Debuting in 1946, the design sold largely unchanged for sixteen years until it was replaced by another Stevens design: the Jeep Wagoneer, the precursor to the SUVs of the last two decades. The Wagoneer remained in production until 1991.

Stevens entered the Studebaker orbit through client connections. He had collaborated with architect James Floria to design the McCulloch Aviation Plant in California. McCulloch, a machine parts manufacturer best known for its small engines, was based in Stevens's home state. He knew both the president of the company, Robert McCulloch, and Sherwood Egbert, a McCulloch vice president who assumed the presidency of Studebaker Motors in 1961.

EGBERT CAME TO South Bend at age 40, confident he could turn the troubled company around. His turnaround plan had little to do with making cars. The board had asked him to diversify the company's holdings to ensure that losses from the automotive division did not bankrupt the company, wanting to eventually transition out of the car business altogether. Egbert refurbished the entire physical plant in South Bend, repainting and cleaning up most of the manufacturing sites. He also reorganized the dealer network. One of his first decisions was to bring back Raymond Loewy and his design studio to create new looks for the 1962 and 1963 cars. He also announced that the company was coming out with a new model in 1964. The sales from the Lark and Hawk models were rising, and Daimler-Benz had committed to giving the company all the Mercedes models it could sell. Egbert had pulled Studebaker into profitability by force of will. His luck did not hold, however. In July 1961, he told the board he had undergone surgery for cancer. After weathering a labor strike and perhaps sensing that he had limited time to put his stamp on the company, Egbert decided to build a new car.

When Egbert came on board, he brought Stevens to South Bend to restyle the 1962 Lark. Stevens's lasting legacy to Studebaker came the same year, when Stevens reconceived Bob Bourke's Hawk as the Gran Turismo Hawk, a swooping extravaganza of a car that built on the design tropes of Loewy's Starliner to create a woozily overwrought luxury car. The interior held the most surprises, including an aircraft-influenced dashboard demarcated into three separately angled sections to give the driver the feeling of being in a cockpit.

Stevens's embellishments to the car's exterior styling maintained Loewy's signature fore-and-aft balance, except for a dazzling—or, depending on the aesthetics of the observer, garish—front end that featured a huge chrome grill meant to mimic the distinctive front ends of Mercedes sedans. Stevens's Hawk was hailed as a triumph of the 1962–63 model year, but the manufacturers were hit with a crippling labor strike the next year, and the brand never recovered.

While Stevens's role as savior of the Studebaker mark lasted less than five years, the success of the Gran Turismo Hawk gave the executives a glimmer of hope that the public would respond to another "halo car" design from the South Bend factory. When Sherwood Egbert decided to create a dynamic sports car to compete with the Corvette, Thunderbird, and smaller sports cars, Stevens would be left out in the cold. Egbert felt the sports car should be a radical departure from Studebaker's previous efforts. Stevens had excelled at improving existing models, but Egbert wanted more. The design for the last best hope of the Studebaker Motor Company would go to Raymond Loewy.

10

Avanti
Car Design Leaps Forward

Sherwood Egbert, formerly chief executive officer for the McCulloch Corporation, a manufacturer of chain saws and other equipment, arrived at Studebaker Motor Company in 1961 ready to pull the troubled car company out of the financial doldrums that had dogged the manufacturer for decades. Supremely confident and carrying the approval of the board of directors, one of his first moves was to bring back designer Raymond Loewy to the company's headquarters in South Bend, Indiana. Egbert asked Loewy to return to perform a facelift on Studebaker's 1962 and 1963 models. Loewy met with Egbert in South Bend on March 9, 1961, and, during the meeting, Loewy received a bigger assignment— one that would deservedly cement his reputation as a visionary automotive designer.

Loewy had parted ways with the automotive company in 1956 amid recriminations from Studebaker's conservative executives that he was introducing "European styling" into the new car models. His hefty contract, with a $1 million price tag, was a matter of some concern to the thrifty midwestern car executives and had caused resentment in the boardroom, on the manufacturing floor, and on the designer's own staff.

Loewy recalled the meeting with Egbert. "He handed me a bunch of clippings about cars which he's been carrying around and asked me if I could do the design in two weeks . . . I did not know the man, but I read him through the sketches he handed me. I knew that Egbert had a natural flair for design. I knew I was working for a man I could respect for his good taste." There is a healthy dollop of flattery in Loewy's description, but the meeting between the two men resulted in the most comfortable relationship between Loewy and a Studebaker executive.

The new CEO, who was described by an associate as having one gear—full speed ahead—had decided the company needed a cutting-edge sports car to compete with the Corvette and some of the many European imports coming into the country. Egbert wrote of his sports car's conception, possibly inspired by Loewy's penchant for the creation story, in the company's annual report, *First Quarter Statement, Ending March 31, 1962*. He bought several sports car magazines, and "by the time I had concluded my trip, I had sketched out an automobile." He was looking for "a single dramatic new model that would bolster both Studebaker's dilapidated image and its sagging morale until the Lark could be redesigned or replaced." Egbert was looking for a "halo car." A halo car, by dint of its "aura" of innovation and excitement, makes the other models in a car line all look better to consumers. A halo Stude-baker might inspire customers besotted by the new car to buy other models. Egbert's strategy was that a prestige sports coupe modeled on the Thunderbird and Corvette, both partly inspired by the 1953 Starliner, would be an impressive complement to the family-oriented Studebaker line and the company's newly franchised import line, Mercedes Benz. Egbert, who knew Loewy had designed a series of one-off sports cars for himself, felt Loewy could design a sports car for Studebaker on a short deadline and make it unique.

He brought in Loewy with a few restrictions. Loewy's designers would be overseen by company styling chief Randall Faurot, giving Egbert, in effect, final say on any of the company's designs. Loewy had a single condition: "Let me do it the way I want to and give me complete freedom of action. I want the authorization to do it my way, far from South Bend . . . I want to be free from interference, and especially free from well-meant suggestions."

EGBERT, WHOSE OUTLOOK on car designers was not colored by ingrained beliefs about "what the buyer wants," was an example of the up-from-the-bootstraps American success story. His father was a pool hall and barbershop owner in Garton, Washington. When one of his father's business investments went bad, the family lost their home and lived in tents for several years. Egbert went to work early in his life and kept working.

He was hired on at MWAK, a multicompany consortium that was contracted to build the Grand Coulee Dam. While there as a laborer, he absorbed so much on-the-job knowledge that he became an engineer. He moved on to Boeing, where he learned more engineering as well as the intricacies of the plant floor. His next career change was to McCulloch, signing on as a management executive.

When Studebaker hired Egbert, his decision to rehire Loewy and his team was not popular with the board of directors, but they backed his request. Unfortunately, shortly after Loewy's return, Egbert was stricken with cancer. The company board tried to take over the company, but Egbert fended off the attempt. Perhaps because he felt he had only a short time to create a legacy, Egbert pushed hard to build the new car.

Egbert wanted a prestige automobile that would appeal beyond the horsepower-and-handling zealots who typically bought sports cars. Egbert wanted the car styled on a five-year cycle, directly rejecting Detroit's policy of yearly updates. Although the five-year design cycle is commonplace in the twenty-first century, Detroit commanded yearly updates of every model line until the 1980s. The Studebaker board approved the sports car project largely because they felt the company needed a car to compete with the Thunderbird. Ironically, the Thunderbird, like the Starliner, was not destined to be a sports car. Ford realized that the car would have limited sales as a two-seater. The company added a back seat to the T-Bird in the late 1950s and, seeing the increased sales of the four-seat Starliner, Detroit carmakers realized that there was a market for a new type of automobile. As families earned more income and women became either part of the workforce or moved to the suburbs, the "family car" became the model of choice for women drivers, leaving men to look for upscale cars that reflected their status in the community. Sherwood Egbert, desperate to debut a new model to energize buyers, felt that Corvette and Thunderbird sales indicated that the market wanted sports cars, not upscale personal luxury cars. To Egbert, Loewy was the only choice to design Studebaker's entry into an already crowded market.

Throughout his life, Loewy often used European custom coach makers to execute his design ideas on existing cars. Aside from the one-off Hupmobile he designed for Hupp using his own funds, Loewy reimagined two Lincolns, two Cadillacs, a Jaguar, several Lancias, and a BMW. The redesigns are more fantastic than serviceable. The redesign of

Raymond Loewy with Studebaker president Sherwood Egbert as they show off the Avanti sports car. Egbert had brought back Loewy and Associates to design a sports car as a last-chance opportunity to turn the failing car company around. Loewy agreed to take the assignment—so long as there was no interference from Studebaker executives. (Raymond Loewy™/® by CMG Worldwide, Inc. / www.RaymondLoewy.com)

the 1940s-era Lincoln Continental, which emphasized the balloon-style body sculpting of the era, is more like a hot-rodder's vision of a luxury car rather than a serious redesign. One addition—two small portholes cut into the rear roof support, later called "opera windows"—would become a staple of Lincoln and Ford designs during the 1970s and '80s. (GM and Chrysler used them as well.) Gossip columnist Dorothy Kilgallen quoted an anonymous admirer as characterizing the Lincoln makeover this way: "it's El Morocco without the rhumba band." Loewy's Jaguar redesign, which sported exaggerated flared fenders, featured a metal exterior "roll bar" integrated into the roofline that Loewy claimed was co-opted by Porsche for its seminal 911 design. The taillights, brake lights, and signal lights were all incorporated into the Jaguar's bumper.

The BMW redesign, completed in the late 1950s, featured twin exhausts extended into the rear bumpers. The bumper featured a sliding "accordion" option that was meant to reduce damage in case of impact. Loewy also designed doors that opened into the roof, which made entry into the car easier and incorporated an all-glass back fastback rear window that provided exceptional visibility.

The Lancia Loraymo (the name is a mashup of "Loewy Raymond") featured a few options that later appeared on the new Studebaker sports car, including an offset hood scoop and flared side panels. It also had a roof-mounted airfoil, which, thankfully, never caught on. There is a photograph of Loewy explaining the airfoil to General Charles de Gaulle at a Paris auto show, and "*mon general* looks distinctly unimpressed." Freelance automotive journalist Len Frank reserved a characteristically venomous review of Loewy's work for the Loraymo. "The bulbous rear was even more clumsily handled [than the BMW], the Italian coach builder having had problems with both shape and method of construction. On the hood was the genesis of asymmetry. The outsized grille was flanked by cutaway flying fenders and headlights on stalks reminiscent of a hammerhead shark." The Loraymo wasn't the eyesore so many journalists claimed it was, but it did serve as a physical manifestation of Loewy's wilder impulses. By exaggerating his personal cars, he could see what worked and what didn't in the real world. It's hard to predict consumer

Raymond and Viola Loewy next to the Lancia Loraymo, a one-off design created by Loewy and executed by freelance car builders. The original Lancia sports car was reconceived by Loewy, who added the off-center air scoop, which was used later in the design of the Avanti. Other improvements, such as the roofline air foil, were less successful. (Raymond Loewy Archive, Courtesy Hagley Museum and Library)

reactions based on drawings and clay models, and having a physical product to gauge acceptance of certain features was invaluable.

To facilitate some of his fantasy cars, in 1956 Loewy hired Pichon et Parat, a coach builder based in Paris. The team was to restyle the body on a new BMW 507 roadster. Bernard Pichon and André Parat ran their custom shop on an abandoned farm outside of Paris. Frank, never wanting for an opinion, disparaged all of Loewy's one-off designing, calling the BMW 507 experiment a body of "surpassing ugliness." The design, a fastback coupe, also looked much like the eventual design for the Avanti. The production model of the 507 was designed by Count Albrecht Graf von Goertz, who worked for the Loewy gang in South Bend after World War II. Loewy never revealed who designed the custom redesign of the 507, but Tom Kellogg, the lead designer on the Avanti team, told *Automotive News* that Bob Andrews, another designer on the Avanti team and the creator of the 1948 "stepdown" Hudson (one of the first cars with slab-sided design), was probably responsible. Kellogg also told the magazine that photos of the BMW were displayed in the makeshift "studio" where the Avanti was created. Loewy owned the unique creation until 1962, when he donated it to the Natural History Museum of Los Angeles County.

Loewy spoke in 1971 of his love of creating cars for his own use in *Industrial Design*: "I was often frustrated working for Detroit, and sometimes by simply observing its designs. Sketching ideas for experimental cars gave me pleasure and released pent-up pressure. Forty-seven years ago, the downslanted hood, the integrated bumper, and the replacement of the grille by an air scoop [near the ground, where it's more effective] were all new ideas. A common device on racing cars, the lowered air intake ventilated the engine compartment most efficiently. The streamlined front and rear ends and the slanted [wedge-like] superstructure were based on aerodynamic principles and adapted for all contemporary automobiles."

During his forced exile from Studebaker, Loewy had kept active in automotive circles. He attended the races at Le Mans every year. He was spotted there one year by Tom McCahill, an automotive critic who described the Loewy sighting in the November 1953 *Mechanix Illustrated*:

Raymond Loewy, Studebaker's designer and chief stylist, proved once again in 1953 that he's the guy the rest of the country's designers wish they were. Back in 1946 he inspired the industry to steal his notch-back Studie designs and in 1953 he came out with a car that made the typical monsters of Detroit look as modern as Ben Hur's chariot in a stock car race. When I was in Le Mans last summer for the famous 24-hour race, I kept stumbling over Loewy in every pit. Unlike some of our home-grown design jackasses, who never stray further from their drafting boards than the nearest saloon, here was Raymond Loewy, checking every new angle and interesting curve [automotive], the products the best car brains of Europe had produced.

Meanwhile, Studebaker's Egbert began a series of acquisitions in automotive-related industries that would lessen the pressure on the car company and also make producing a sports car more economically feasible. He bought out Chemical Compounds Inc., which manufactured the engine additive STP. He spent $275,000 (approximately $2.2 million today) for Paxton, which developed an automotive supercharger, in March 1962. Both acquisitions would prove to be critical for the new sports car's success.

As Egbert set the stage for interrelated manufacturing, Loewy brought together his own team of car designers. Loewy, elated at the return to Studebaker, flew home to Palm Springs, California, where he lived in a modernist home. But he rented another house in Palm Springs, which was about as far from South Bend as he could get, and assembled a design team. The small tract home was what author Tom Wolfe would later call a "ranchburger." A one-story bunker with an attached carport, the house was spare, but then again Loewy wasn't living there. The team he hired was. He asked John Ebstein, a native of Germany who had worked at Loewy's design firm as an airbrush artist and photographer, to be the head of the design team. Tom Kellogg and Bob Andrews, both car designers for Loewy, were asked to join the project. Andrews and Kellogg were in charge of sketches. At the Palm Springs design house, Loewy had taped up photos of several cars that would serve as influences for the new sports coupe. The 1961 Lincoln Continental was included for its distinctive thin-bladed quarter panels, and the Jaguar XKE was celebrated for is sinuous ground-hugging appearance. Loewy also included a rough sketch of a car with a "Coca-Cola curve." The Coca-Cola curve was Loewy-speak for a car that had a pinched waist that flowed inward from the hood and trunk. The designer constantly returned to the

A rare candid photo of Raymond Loewy in the Studebaker design studio in South Bend, Indiana. Note the car sketches on the wall. It was common practice to save just the final sketches for the client, making it difficult to trace designers' work throughout the process from idea to final product. (Raymond Loewy™/® by CMG Worldwide, Inc. / www.RaymondLoewy.com)

Coca-Cola reference he first broached in the 1949 *Life* interview where he praised the bottle's "Callipygian curve," but Jaguars and Ferraris also featured subtle hourglass shaping, and it seems clear from his writing and from the recollections of the design team that Loewy was inspired to see the shape translated into sheet metal. Many anti-Loewy car enthusiasts feel this is yet another instance of credit-grabbing for his hazy claims to designing the Coke bottle.

The team assembled for the Avanti project was sequestered for the duration of the job. Bob Andrews described it as being "like a cloak and dagger movie. We had no idea what was up except that it was terribly secret and we'd have to develop the thing within an untypically short amount of time. Once we got there R.L. closed us up tight. He wouldn't even let us out for a night on the town. He disconnected the phone, stopped all the clocks and banned wives and girlfriends. We worked sixteen hours a day, every day, for weeks."

Andrews felt his time working for Loewy, no matter how long the hours, were the finest years of his career. "I had a very good relationship with him. One of the reasons . . . was that I never really tried to get to know him too well on a personal level. I'd seen that happen. It was disastrous for some people . . . Mr. Loewy was such a fine director that when you worked on a project for him, it was the most important project going on, not only in the United States, but in Europe, at that time. . . . For Loewy to extend that to me, I felt just great about it and did some of my best work for him."

After taking inspiration from the preliminary sketches, the team immediately started work on a one-eighth-scale clay model and finished it within the two-week deadline. Egbert flew down to the makeshift design studio and approved the design, whereupon Loewy shipped the entire operation to South Bend to create a full-scale model. Known as the X model, or the X-SHE (Egbert's initials), it was clear to executives this car was unlike any previous Studebaker model. In fact, the new design was unlike any previous car design. The prototype was called the Avanti, Italian for "forward." Bob Andrews credited D'Arcy Advertising Agency with naming the car, but Loewy always gave Egbert credit for the name. Records bear out Andrews on this point.

Andrews recalled that Egbert did not impose his views on the design process, or at least not as much as other executives. "Egbert did not try to become an automobile designer overnight, which happens so much.

Raymond Loewy, dressed in the "casual" western garb he preferred when in residence in Palm Springs, poses in the "ranchburger" home he rented for the Studebaker Avanti design team. (Raymond Loewy™/® by CMG Worldwide, Inc./www .RaymondLoewy.com)

When a new executive officer comes in, he thinks he's got to know, and [Egbert] did not try to know . . . So the usual things that would have happened in engineering to cheapen that car—to screw it up—did not happen for that reason."

Meanwhile, at General Motors, William L. Mitchell was revolutionizing the look of the company's vast fleet, focusing most of his personal attention on a series of Corvette sports coupes that would serve as a Holy Grail of sorts for Studebaker executives and for Egbert. Chief of GM design from 1958 to 1977, Mitchell directed the look of more than 70 million cars. Like his mentor Harley Earl, Mitchell did not personally design cars from start to finish, but instead acted more "like a producer of films than a director." Earl's designs were rococo, whereas

Mitchell retained the GM policy of five-year evolutionary designs but stripped the car's look to its essence. According to automotive journalist Phil Patton, Mitchell liked to call his design style "London tailoring," by which he meant removing the tail fins, pumped-up bumpers, and hulking bodies of the 1950s in favor of a sleekness that integrated the auto body into a continuous design. "I wanted to put the crease in the trousers," Mitchell said.

Mitchell replaced Harley Earl just as GM's models were being forsaken for the elegant, faster Chryslers designed by Virgil Exner. Exner took the tail fins Earl favored and incorporated the Forward Look into the rakishly angled fins, grille, and body. GM designers saw that the company's planned 1958 models were pallid, safe designs compared to the modern looks created by Exner and threw out the entire 1958 design inventory while Earl was on vacation. When he returned, Earl acknowledged that the existing designs had been lifeless, and Mitchell was soon made Earl's heir apparent. Upon his retirement, Earl shared his unique philosophy for design: "My primary purpose has been to lengthen and lower the American automobile, at times in reality and always at least in appearance. Why? Because my sense of proportion tells me that oblongs are more attractive than squares, just as a ranch house is more attractive than a square, three-story flat-roofed house or a greyhound is more graceful than an English bulldog."

Mitchell, whose own personality and habits—rife with hedonistic and heedless pastimes and prejudices—could be compared to those of a bulldog, replaced the antiquated system of designers reacting to the profane commands of one designer. He decreed a simplified system where small drawings were blown up into wall-size images that could be easily rated and edited by design teams. At the Loewy shop, the final decision came down to the owner, but Bourke shepherded most of the designs through the process. Mitchell was more open to ideas by junior designers as well. His opinion of his own taste was sure, though, as was his opinion of other members of the GM team. "Goddam right I'm a dictator. If I don't have opinions I shouldn't have my job. . . . Frank Lloyd Wright didn't ring doorbells asking what kind of houses people wanted . . . How the hell can you explain a martini to anybody. I don't like engineers, salespeople or anybody. I don't give a shit what they say," he said.

Ron Hill, a former GM designer best known for the Italian-influenced late '60s Chevrolet Corvair, considered Mitchell much more approachable, saying, "Earl was much more pugnacious. Mitchell was a better designer and a more tasteful individual." Mitchell promoted many young designers, including Hill, Chuck Jordan, Bill Porter, and Irv Rybicki. Dubbed the "Young Turks," these stylists defined the automotive style into the 1980s. They pushed big, aggressive designs, which eventually morphed into smaller, aggressive "muscle cars" that used horsepower as a selling point—the design was secondary.

The huge talent pool at GM set the market in almost every product category. Although other cars manufactured by Ford and Chrysler could occasionally break GM's stranglehold, it was difficult to create a car that would stand out in the parking lot compared to the Corvette, the Buick Riviera, and Cadillacs. Other companies had to push beyond the boundaries of the Big Three designs to get noticed. In design and execution, the Avanti was the car that looked unlike anything on the road. Nearly simultaneously in Detroit, three Ford design teams were slaving over a commissioned design for an inexpensive sports car. Like the more expensive Avanti, the Ford design was based on an existing chassis, but the engineers moved the seat position back to accommodate a longer hood. The car, based on the Ford Falcon chassis, was the Mustang. Debuting in 1964, it was created by Ford chief designer Joe Oros and stylist John Najjar, with contributions from other staff designers, including L. David Ash and Gale Halderman.

The Mustang still had more chrome than any of Studebaker's cars, but the ultimate design forever changed the visual look of Detroit cars. Similarly, the Avanti design team did not really create the sports car out of thin air. Unlike most design studios and truly unlike the pristine, spacious auto design centers of Detroit, the cramped ranch house "studio" of the Loewy team dictated that workspaces be spread throughout the house, Kellogg near the fireplace and Andrews on the kitchen counter. Loewy was immersed far more in the design of the sports coupe than he had been previously, and the lion's share of the design work was done by Ebstein's team. Kellogg, the main sketch artist, shipped cascades of sketches to South Bend, where Studebaker chief engineer Gene Hardig had to create the car. Most of the team's design was approved, except for Loewy's initial dual-headlights design, which was eliminated in favor of single headlights for cost.

John Ebstein had the longest service for the Studebaker account. Although many of Loewy's designers were expected to produce drawing in a variety of styles, from pastels to pencil to ink to charcoal, Ebstein was a master of the airbrush. The painstaking technique was well suited to depicting the gleaming surfaces of trains, aircraft, locomotives, and ships. Ebstein was rarely assigned to design accounts until Loewy asked him to join the Avanti design team. Ebstein saw the assignment to create the Avanti as his chance to leave a legacy in automotive design.

Tom Kellogg, who provided most of the conceptualization for the Avanti, graduated from the Art Center College of Design in 1955 after serving a design internship at Ford Advance Styling. He grew up on an Illinois farm, and one of his cherished memories was learning to drive sitting on his father's lap in an old Studebaker. He was recruited by Loewy and Associates to work on the Avanti and stayed through 1963. He left in 1964 to open his own design firm in Newport Beach, California, and went on to design exteriors for Rolls Royce and Porsche, motor homes, and several lines of fiberglass boats. He created designs for Wedgwood china, designed the interior of the McDonnell-Douglas DC-10, and most impressively, at least for pop culture fans, the shuttlecraft for the original *Star Trek* television series.

Kellogg's personal letters from 1965 and 1966 put to rest the perception among automotive critics that working for Raymond Loewy was an onerous and soul-draining experience. He tells Loewy he heard that the company is considering opening a Los Angeles office and asks to be considered for reemployment. He wrote a subsequent letter as well, outlining his current work and work for Loewy. A clue to Loewy's impression of Kellogg is found in Loewy's archives from a 1970 document pertaining to the design of the Avanti. "He wanted to try all sorts of things. To keep Tom with his eye on the ball was no cinch."

The designers and Studebaker executives went back and forth on whether to design the Avanti as a two-door two-seater coupe, or as a four-seater. Egbert wanted a two-seater. Loewy and engineer Eugene Hardig fought for four seats. Eventually, it was decided to pursue both designs. In another design mini-battle, Egbert fought for no sun visors, while nearly everyone else on the team insisted the car should have standard sun visors. Loewy won, but he was asked to design razor-thin visors that were practically nonfunctional. Loewy's most important contributions were suggestions to eliminate a conventional grille in favor of a "chin scoop" slung under the bumper, which made the front profile of the Avanti look like no car manufactured before (or since, for that matter), and a decision to use razor-thin slab quarter panels. Loewy and Kellogg also came up with an off-center bump or "blister on the bonnet" running up the hood. The bump supposedly helped direct the driver's vision forward, but in retrospect it probably made the car's visual signature more unique. Loewy said of the bump, "It made the car and driver integral, like the gunsight of a gun."

Studebaker's Hardig, who had worked almost seamlessly with the Loewy shop on engineering the Champion and Starliner designs, also took on the Avanti build-out assignment. In addition to the cutting-edge design, the new executive management team was somewhat more open to European styling under Egbert's stewardship. They could see Triumphs, Jaguars, and Mercedes edging onto the nation's highways. Loewy also asked Hardig to stiffen the suspension and make the steering more responsive, a 180-degree reversal from the cushiony suspensions and sloppy steering that emerged from the Big Three plants. The Avanti had a minimum of decoration. Loewy wrote about his "chrome policy" in a 1950 *Science and Mechanics* article titled "Car of Tomorrow":

> Chrome areas of tomorrow's car will be reduced, but probably never will be eliminated entirely. Nor is this absolutely desirable in all instances, since time and time again the public has proved that it wants chromium. If it isn't on the car, buyers will add chromium in the form of accessories and extra gadgets of poor design. So it is a better idea for the designer to handle the chrome himself, and do it well. He can reduce the number of meshes, openings, lacework patterns, rods and cross-hatches, grilles and bars. Chrome will be used in simpler, plainer masses, not over-styled with gingerbread, or in designer's language, "schmaltz."

The Avanti's most innovative design feature was its wedge-shaped front where the grille would customarily be. The single headlights were recessed but outfitted with a clear plastic cover that integrated into the wedge-front design. The brochure for the Avanti, designed by the d'Arcy Advertising Agency with a silver-on-blue motif, emphasized its "Aerodynamic Wedge Design!" The copy is all Loewy-speak: "Its clean crisp lines represent a fresh departure from today's trend toward over-ornamentation . . . Its wind-resistant wedge shape holds the body snug to the highway for better reliability—easier, smoother cornering. Here is

America's most advanced automobile built in the Studebaker tradition of engineering excellence and quality craftsmanship."

The interior of the car was innovative as well. The seats, minimally padded, were upholstered with Textileather, a brand name for upholstery now lost to the ages. The dashboard was based on aircraft control panels. The center control panel was located between the bucket seats. All the controls were illuminated with red lamps for better visibility at night. In addition, Loewy added rocker-style switches for the lighting and fan controls on a console mounted on the roof, another touch inspired by aircraft. The dashboard had "crash padding" (which was partly responsible for relocating some of the controls to the upper console) that was hardly enough to protect a helmeted driver but more than other cars were offering at the time. The car also featured a reinforced roll bar, called a "padded safety arch" in the car's advertising, and reinforced windshield. The padded safety arch was a modernist design. Instead of hiding the roll bar within the roof structure, the feature was emphasized, protruding in a graceful arch from door to door.

One of the most interesting options was the Avanti Beauty Vanity. From the brochure: "This will surely interest the ladies. Within the large glove compartment is the convenient, self-illuminating Avanti Beauty Vanity Tray, complete with pop-up mirror for that last-minute primping before stepping out." For those who enjoyed stepping out with something larger than a lipstick, the Avanti also featured an extra-large trunk, something other sports cars could not compete with. The designers gained the extra roominess by extending the trunk beneath the back window. Compared to the postage stamp–size trunks on the early Corvettes and European sports cars, the roomy trunk of the Avanti was another selling point. The trunk also featured an access door through the back-window shelf, "to reach those indispensable items which clutter up a car's interior."

Some of the Avanti's materials were not as groundbreaking, and in fact were responsible for nagging problems that the model could never overcome. Although fiberglass had been previously used for body moldings on the Corvette, Studebaker chose the material to save costs and manufacturing time. The individual car body would be more expensive, but factory-tooling costs would be less. Studebaker even marketed the fiberglass construction as a noise suppressant: "Makes the AVANTI quieter!" was the tagline in the brochure. Engineers at the company went to

Ohio Molded Fiberglass, the same company that supplied body materials for the Corvette. Nearly as soon as the contract was signed in 1962, Molded Fiberglass went on strike, so Studebaker decided to start its own fiberglass operation in South Bend. Eventually, the Avanti bodies would be assembled at production lines in both South Bend and at Molded Fiberglass's plant in Ashtabula, Ohio.

Mechanically, Gene Hardig created one of the more advanced production cars for the time. Using the frame of a Lark convertible, he outfitted the Avanti with front and rear anti-sway bars and rear radius rods. Originally, the car was to feature an independent rear suspension, but that was eliminated early in the design in favor of front coil springs and rear leaf springs. Heavy-duty shock absorbers were added as well. He asked Borg Warner to engineer an unusual transmission that operated manually in first gear and slipped automatically into second and third gear from the drive position.

The Avanti's Bendix disc brakes were the first caliper disc brakes used as standard equipment in an American production car. The engine was a 289-cubic-inch V-8. The Studebaker marketing literature cemented yet another aircraft connection by calling the engine a "Jet Thrust V-8." The company made an overt pitch to the luxury car market, highlighting the many engine parts with chrome, including the rocker arm covers, dipstick, oil filler cap, voltage regulator, and air filter. The Avanti also offered, on some models, a Paxton supercharger. One experimental version featured two superchargers and fuel injection. The twin-rail ladder chassis was Studebaker's standard sedan chassis, developed in 1953. Despite Hardig's use of stabilization and anti-sway bars, the suspension was not much improved.

Once the car was in production, the company proceeded to mount a full public relations blitz to Studebaker dealers, salesmen, and media. The Avanti was loaded into a "boxcar" aircraft and flown nearly 15,000 miles on a tour of dealerships, car shows, and special appearances. The new car was seen by more than seven thousand employees, independent dealers, and reporters. The company produced brochures encouraging independent dealers to take on Studebaker or at least the new Avanti. Calling the car a "hot, saleable line," the company produced material such as a brochure titled "Here Are 35 Solid Reasons Why It's Easy to Sell This Line." The company cranked out more publicity materials than previous Studebaker makes, including safety-related ads touting the

Avanti's brakes and a giveaway contest that offered 350 automobiles to lucky contestants.

In 1962, Andy Granatelli broke twenty-nine speed records on the Bonneville Salt Flats driving an R3 (one supercharger) Avanti, making the new sports car faster than any other American stock car. Len Frank, a writer for *Motor Trend* and *Popular Mechanics*, in an article excoriating Loewy, had high praise for the final product: "Too often a dream car is turned into a nightmare by those of little courage and less taste. The Avanti was spared."

When the Avanti debuted, the *New York Times* had this to say: "There is some concern among Studebaker's conservative engineers who feel that the avant-garde vehicle might hurt the company principally identified with the highly conventional economy Lark." After the Avanti's opening splash, Egbert initiated plans to debut two more model lines and tasked Loewy to work on designs that would extend the Avanti look to the new cars. At this point in Studebaker's corporate history, inconsistency in model designs was practically a corporate credo.

Two Studebaker Avantis parked at Indianapolis Motor Speedway in October 1962. *Left*, Studebaker CEO Sherwood Egbert. *Right*, Andy Granatelli, a famous automotive engineer who helped Studebaker set land speed records in a supercharged Avanti. Granatelli later became CEO and spokesman for STP oil additive. (From the Collection of the Studebaker National Museum, South Bend, Indiana)

After various administrations added on to Loewy designs or brought in other design voices, it was fortunate Studebaker models had even a few design tropes.

Glowing reviews by advance auto critics and other media couldn't energize Studebaker to get enough Avantis into production. The original fiberglass manufacturer, Ohio Molded Fiberglass, had production problems due to striking workers, and the bodies produced before the strike tended to discolor in the sun. The quality-control problems forced Studebaker to build its own body facility in South Bend. Otto Klausmeyer, assistant manufacturing manager for Studebaker, recalled the problems associated with Molded Fiberglass: "It was catastrophic. The doors wouldn't close, the hoods were out of line, the fender contours were mismatched. When we tried to drop the rear window into position, it fell through the hole!" By the time construction for the South Bend fiberglass facility was finished, the market had moved. Most of the car enthusiasts who were looking to buy the latest plaything had grown tired of waiting for delivery and decided to buy Corvettes or imported sports cars.

The Avanti featured a fiberglass body that incorporated a "pinched-waist" curve and a groundbreaking rear body style later copied ad infinitum by car designers everywhere. The original rear windows sometimes flew out of the frame when the car reached cruising speed. (Raymond Loewy Archive, Courtesy Hagley Museum and Library)

The company also indulged in some pop culture marketing. The Avanti was featured as a major prize on the game show *The Price Is Right*. Avantis were purchased by entertainer (and, later, sausage magnate) Jimmy Dean, comic Dick Van Dyke, All-Star pitcher Sandy Koufax, and late-night television host Johnny Carson. Frank Sinatra owned an Avanti, as did television writer Rod Serling. Television humorist Herb Shriner was killed in an auto accident driving his Avanti.

Reactions in the automotive press were positive. Tom McCahill, writing in the September 1962 issue of *Mechanix Illustrated*, wrote, "If this goes over as Studebaker hopes, the city of South Bend should erect a statue of Raymond Loewy right in front of city hall." Alas, it was not to be. The four-seat design proved that there was a market for a sporty car with extra room, but it was not a market that Studebaker could take advantage of. Two years later, the Mustang was introduced to monumental sales. Studebaker historian Thomas Bonsall believes the Avanti influenced the Mustang design. He pointed out both cars were built on existing platforms and both had similar bumper styles. Only *Road and Track* was disdainful of the Avanti, calling it "a car for the driving sport as opposed to the sporting driver."

THE AVANTI WAS THE FINAL DESIGN that Raymond Loewy executed for the Studebaker Motor Company. It also was the single car design that Loewy personally supervised. The car designed for an all-fiberglass body meant that 130 parts had to be glued together for each model, which posed a quality nightmare. In addition, the all-plastic body shrank 2 to 3 percent, causing major maintenance headaches, not the least of which was the tendency to have its rear windows pop out at high speeds. The quality-control problems kept the popular car from truly taking off. Egbert had promised dealers that he would deliver 1,200 Avantis in one month, but the company managed to produce just 1,000 for all of 1962. In all, Studebaker churned out 4,647 Avantis in eighteen months of production between June 1962 and December 1963. The car, which was never intended to singlehandedly save Studebaker, failed in its real purpose, which was to lure buyers into the dealerships.

Loewy, in remarks before the Society of Automotive Engineers, said, "no one can expose a body to the public, arouse excitement over its form and design, then deliver nothing for 10 months—except possibly Brigitte Bardot. Avanti sales accelerated from zero to zero in less than

12 months." Designer Andrews was more specific about why the Avanti failed. "By the time the Avanti came out, there were very few dealerships left in the country to sell them. We had a lot of back-alley service station dealers that didn't even have room for one car . . . These were very slim times and they'd lost a lot of good dealerships. So here we had a $5,000 [about $40,300 today] car, and we had a bunch of dealers that didn't even know anyone that had $5,000 for a car."

The obstacles of competing with the Big Three were too big to overcome for Studebaker. Little wonder. The General Motors Technical Center, designed by Eero and Eliel Saarinen under the watchful eye of Harley Earl, opened in 1956 in Warren, Michigan, and dwarfed any other automotive design center on the planet. More than a thousand stylists and technicians worked there, as well as a full executive staff. The plan featured a 22-acre lake outfitted with a huge central fountain and a smaller fountain designed by Alexander Calder. There were twenty-five buildings on 325 acres, most conceived in steel, glass, and aluminum. The domed roof of the Styling Auditorium is still one of the most striking modern buildings in the United States.

Loewy's offices in South Bend were efficient, but he was continually playing a different game than the Big Three. The sheer size of the Detroit carmakers' design facilities meant that Studebaker and other independents were always two steps behind. In the GM studios, Earl's office faced the lake, full decorated with overstuffed furniture that looked remarkably like the "balloon" cars he loved so much. And throughout his executive career he used his offices as theatrical sets. His original offices were laid out in a long, hallway-like room in which designers had to walk a long way toward his desk. As they arrived at the desk, Earl's chair platform rose slightly so that Earl remained at eye level with anyone standing opposite. His offices in the Design Center were more visitor-friendly, yet still large, so the experience was more like visiting an international potentate. Earl's successor, Bill Mitchell, used the grandiosity of the design studios to similar effect during his successful tenure. Studebaker, with a small-scale operation run by an outside designer, could not commit the same resources, neither to Loewy's office nor to the car production line.

That Loewy and his team were able to create a unique production car in a matter of weeks at least proved that great design could be done anywhere, even in a Palm Springs ranchburger. The three-person design team under Loewy created a design of true originality in the Avanti,

no doubt accounting for its long life as a custom sports car after Studebaker's demise. Many Hollywood directors used the Avanti as shorthand
for forward design. The film *Gattaca*, set several centuries into the future,
features a scene in which the hero crosses a busy thoroughfare. The cars
float by, suspended on air; many are Avantis and Starliners. The Avanti
contains all the design intelligence Loewy and his team accumulated
over thirty years. As designer Bruno Sacco writes, "He was never a man
to take small steps. It was in his nature to jump several squares at a time
on the checkerboard. And, as his association with Studebaker shows,
he managed to carry it off, and left his mark on automobile design."
Loewy never tired of his crowning automotive achievement, noting in
Industrial Design: "I still keep two beige Avantis, one in Paris, one in
Palm Springs."

He would later write in his archives that "The two great experiences
of my professional career were Avanti and the work we are now doing as
a consultant to NASA." Avanti's weakened sales figures were one of the
final elements in Studebaker's demise. In fact, the company closed its
South Bend plant shortly after Avanti production ceased in 1963. Studebaker executives, faced with declining interest in its entire product line,
saw no need to retain Loewy as a consultant. Still, Loewy never got over
the final break with Studebaker. After being let go in the 1950s, Loewy
barely mentioned the company in *Never Leave Well Enough Alone*, and
in *Industrial Design* he bade Studebaker ill wishes. "My decades with
the company were exhilarating and unforgettable, and my respect for its
engineering department immense. I leave it to others to uncover the reasons why such a great, prestigious firm, having at last found its market,
finally disappeared at a time when it was admired throughout the world
and when the Avanti had just come out with a backlog of orders. It was
an industrial tragedy."

IN OCTOBER 1963, Sherwood Egbert entered the hospital to undergo
further surgery for his cancer. He soon stepped down on indefinite leave
and with him went Studebaker's enthusiasm for remaining in the automotive industry. Shortly after the debut of the Avanti, on November
25, 1963, the production plant in South Bend closed its doors, laying off
6,000 workers and leaving an eighty-six-day supply of unsold cars in
South Bend. The eight other Studebaker divisions were all in the black
when the doors were shut.

Byers Burlingame, Studebaker's new president, made the decision to continue car production in Hamilton, Ontario, on the St. Lawrence Seaway near Buffalo, New York. Auto analysts saw the move as a last-ditch effort to save the company. In a press conference, Burlingame claimed the Canadian plant could turn out 30,000 units per year and make more money owing to lower labor costs. At the time of its closing, Studebaker was the oldest surviving nameplate in the car industry.

In 1963, the company heavily promoted its Lark compact. But the compact push came too late, as the market had moved to larger, higher-horsepower "muscle cars." In addition, the cash-poor firm could not afford to pay for the tools and dies required for a major style make-over. In 1963 prices, a small tool changeover at any car company cost $30 million, and a major retooling was budgeted at $300 million. Such costs were well beyond Studebaker's means. Production problems had doomed the Avanti, and sales figures for its other models could not wrench the company out of its death spiral. Dealers blamed the Avanti for overall poor sales, charging that the new car overshadowed the plain-vanilla styling of the rest of the Studebaker product line. The South Bend plant, today a museum dedicated to Studebaker cars and the company's history, shut its doors in June 1963.

Many analysts, including one in a *New York Times* article, pointed to Studebaker's lack of a strong dealer organization. The article cited a trade survey in Detroit, which said that Studebaker dealers were considered "a complacent and non-competitive group."

The final indignity at Studebaker was when the 6,000 employees who remained on the payroll when the factory closed discovered that the company had gutted their pension plan. The company plan had offered pensions of $80 or $90 per month to retired employees. But the legacy of the Studebaker family and its concern for worker health and welfare had long since been pushed aside by financial men interested only in the bottom line. The sudden closing gave the company the opportunity to cut the pension plan, giving retirees and retirement-eligible employees full pensions and offering vested employees under age 60 about 15 percent of the value of their pensions. The company denied payments to any non-vested employee under age 40. Even if the company had wanted to offer better pensions, there was no cash in the till. Executives had funded business acquisitions using pension funds. The ensuing negative publicity resulted in Congress passing the Employee

Retirement Income Security Act of 1975, which required companies to establish minimum standards for pensions and to purchase insurance to protect pension assets.

More than 22,000 workers were employed in South Bend at Studebaker's peak era. When the plant closed in 1963, the average age of the employee roster was 54, and few had ever worked elsewhere. An internal personnel memo from S. T. Skrentay to C. T. Gallagher perfectly captured the desolation the town and its citizens felt: "The possibility that the South Bend operation could collapse was a chunk of reality that was never allowed to creep into the employee's mind. It never figured into his planning. If a mood permeates the workforce—it is the mood of despair and disillusionment. It is also a mood of frustration, a frustration born of the fact that human experience had not prepared the average employee—especially the older worker whose future appears particularly bleak—with the tools to cope with the catastrophe he now sees as his future."

When the company closed, freelance designer Brooks Stevens blamed Egbert, Loewy, and the Avanti. "The public didn't understand that car," Stevens was quoted as saying in a newspaper interview. "I wish now we had that five million for our family cars. We might still be in business." Unaware or oblivious to the company's executive manifesto to phase Studebaker out of cars entirely, Stevens quixotically continued submitting design ideas to Studebaker executives. One was for a concept car to be shown at an auto show. Calling the car a "Mercedebaker," Stevens created a throwback design based on the roaring twenties classic the Mercedes SSK Roadster. This retro design proved so popular that Stevens's sons, Steve and David, started their own custom car company, naming the flagship model the Excalibur. Mostly sold to professional athletes and Hollywood celebrities, the Excalibur was manufactured under the Stevens family until 1986.

The Avanti was to have a much longer life beyond Loewy and Studebaker. Shortly after the Studebaker plant closed its doors, two South Bend car dealers, Leo Newman and Nathan Altman, bought the rights to produce the Avanti as a custom, high-end sports car. The car was produced on a reduced-scale assembly line and sold in its first incarnation into the 1980s, retailing for about $30,000. They bought six buildings in the company complex and hired company engineer Eugene Hardig, who advised them to hand-build about two hundred cars per year aimed at the luxury car market. Dubbed the Avanti II, the custom car was manu-factured from 1965 until 1983. It was profitable every year until 1983. The cars were initially sold through the old dealership network, but eventu-ally the Avantis were purchased directly from the factory.

Several other entrepreneurs and car enthusiasts took over the company, including Steve Blake; Michael Kelly, a South Bend entre-preneur and car dealer; and, later, in partnership with Kelly, Jim "J. J." Cafaro. Cafaro introduced a four-door, grand-touring sedan version of the car, outraging purists. He also proposed a limousine version of the car. Cafaro bought out Kelly in 1988. The various incarnations of the post-Studebaker Avanti went through periods of boom and bust, some caused by overambitious owners, others caused by the widespread use of an existing GM chassis as the basis of the car. The unpredictability of this chassis, which like the chassis of all the Big Three, was subject to change or cancellation, made it difficult to maintain production stabil-ity over time.

By 1990, the Avanti had evolved into a high-end luxury tourer with digital cockpit and an in-cabin television set in the middle of the back seat. Through a series of buyouts and bankruptcies for the Avanti II businesses, Millersville, Pennsylvania, adman Jim Bunting decided to form a new company to create an entirely new Avanti prototype in 1995. Bunting, who dubbed the new company AVX, for Avanti Experimental, contacted Tom Kellogg, a member of the original design team, to create a new prototype. Kellogg used a 1993 Pontiac Firebird as the model for the new car. Bunting was unable to manufacture but a few cars. South Bend businessman Mike Kelly returned to rescue the Avanti line in 2001 and eventually moved production to Cancun, Mexico. The company ceased production in 2006.

STUDEBAKER FAILED FOR A variety of reasons, but it lasted more than a hundred years as a transportation company and was one of the last independent car companies to close its doors. The ultimate reason for its downfall can be traced to another product introduction, one that occurred a decade before the Avanti. The company experienced quality-control problems in 1953–54, but the severe drop in sales was more likely due to inept designs for the family sedans. An even more pointed factor in its demise was the monolithic competition posed by the Big Three. They could create and dominate new markets, and, more importantly, they could weather horrible markets, as Ford did in the 1920s and GM did in the 1970s and '80s. The Big Three also had lower unit costs. It cost Studebaker more to design a new car, essentially because Ford and GM bought most of their parts from subsidiaries. Studebaker could not compete in the full-line economy-family-luxury market. Former president Erskine came close, but the Depression effectively erased the opportunity. When the 1953 models were tanking in the marketplace, the company's massive payroll, coupled with the low productivity of the plant, meant that the company had to sell an absurdly high volume of cars to make a profit.

The 1953 models, particularly the Starliners, were the death knell of the company. The smaller, lighter cars offered less chrome when the market was going 180 degrees the opposite way. The design for the sedans was poor, and the coupe and sedan lines did not share common parts. The company could not earmark millions to retool the entire product line, like Chrysler did in the late 1950s for the Forward Look years. As a result, Studebaker lost two-thirds of its market share in three years. Their move into the compact car market was too little, too late. In addition, the company abandoned its truck market at almost the exact moment in time when the national truck sales were taking off. On top of all these mistimed moves, the company was losing dealerships. At the end of its corporate life, Studebaker had 1,200 dealers while GM had 14,000. Car dealerships depended on new car sales, service, and used car sales for profitability. Big Three dealerships could weather bad times by depending on profitability from two of these centers, but independents had to wring profits from all three. Studebaker dealers, if they were successful, often transitioned into GM, Ford, or Chrysler dealerships when an opportunity arose.

Perhaps the most shortsighted strategy pursued by Studebaker's executives was its refusal to reinvest profits into its core business. The excess cash brought in by successful models such as the Starliner and Lark was directed into dividends, bonuses, and increasingly fat union contracts for workers rated at the bottom of the industry for productivity. Frederic Donner, chairman of the board for GM from 1958 to 1967, zeroed in directly on Studebaker's shortcomings. "Did you ever stop to wonder what they did with the lush profits of the war years? If they reinvested them in the business? We didn't drive them to their present condition. They drove themselves there."

As for Loewy, he never stopped belittling American cars. In 1979, he was dismissive of an American auto industry that, to be kind, was at its nadir in design creativity. General Motors and Ford, buffeted by a public outcry for fuel-efficient, smaller cars, were taking existing designs, which were still using the slab-sided design pioneered by Bill Mitchell in 1963, and shrinking them to fit smaller cars. The outcome was such forgettable designs as the Ford Granada as well as the worst of the various versions of the Thunderbird, Cadillac Cimarron, and many others. Loewy went on to compare his Avanti's sleek door with the behemoth doors favored by Detroit. "In this age of fuel shortages you must eliminate weight. Who needs grilles? Grilles I always associate with sewers. I'd also kill chrome forever, or any other applied junk."

Thomas Bonsall succinctly summarized the company's demise. "The fall of Studebaker was not inevitable. It was the result of peer decisions in South Bend, hastened by opportunities missed. It was a tragedy that need not have happened and in that may lie the greatest tragedy of all."

11

Becoming a Businessman
Building an Industry

L oewy stressed that a product design emerges through teamwork and "unavoidable interferences"—comments, critiques, and technical specifications. He warns that the design can be compromised at any stage by any combination of those factors. "To protect his infant during this tumultuous Northwest Passage across the wilderness of research labs, mock-up rooms, model shops, conference boards, and testing labs, the designer must be gifted with the vigilance of the trapper, the diplomacy of Talleyrand, and the perseverance of a Columbus."

Loewy wrote the preceding thought in *Never Leave Well Enough Alone*, and once the reader cuts through the author's natural tendency for hyperbole, what's left is a perfect description of how the industrial designer works. Loewy's autobiography differs from most books written by designers in that he did not use his pages to give manifestos that explained or extolled his aesthetic. Instead, Loewy's aim was legend-making. In *Designing for People*, Henry Dreyfuss devotes the premium amount of space to explain ergonomics, the study of people's efficiency in their working environment. Using two stand-in mannequins that Dreyfuss dubbed "Joe and Josephine," he details how to design products that best fit the human lifestyle. The same goes for other designer/authors, such as Walter Dorwin Teague's *Design This Day*, Harold Van Doren's *Industrial Design: A Practical Guide*, and Bel Geddes's *Horizons*.

Loewy, by contrast, spends a great deal of time discussing not how to design products, but rather how to run a design business. He dedicates two full chapters to telling readers how to organize an operation, while sprinkling tidbits of management wisdom throughout the rest of the book. *Never Leave Well Enough Alone* could be a precursor for the modern executive advice book.

Loewy starts his explanation of the design business by introducing the client, an ice cream freezer manufacturer called, with Loewy's penchant for wincing puns, the Nadir Company. The decisions he describes make it clear that Nadir is a stand-in for his experiences with Sears and Frigidaire. Loewy discusses the client's strengths and weaknesses, including poor packaging, bad wrappers and packing boxes, and ineffectual point-of-sale displays and sales literature. Loewy suggests that readers follow up on the client's research with research of their own. Acting as an urbane ringmaster, Loewy describes a meeting with Nadir's sales manager in which the bumbling sales representative is constantly upstaged by Loewy's design team. The fictional presentation includes a live examination of the Nadir freezer and its main competition. Loewy gives the client's copy middling reviews and asks the company's man if he can lighten the final design by 4 pounds.

Loewy assigns a team to visit the company's plant, where his people estimate how the new designs might fit into the manufacturing process. They also are asked to complete a study to determine whether investment in new manufacturing equipment for the designs would be cost-effective. Blueprints are obtained from the manufacturer, and many sketches later, the design team has made two versions of a redesign—all on paper. The various sketches are winnowed to about four, and calls are made to Nadir to determine whether the designs are feasible for manufacture. The final four designs are chosen. Four separate full-size clay models are made because, according to Loewy, home freezers are small enough to reproduce in real dimensions.

Loewy describes the design presentation as if it were a Hollywood premiere. Spotlights are trained on the models, and chairs are arranged in a semicircle around the models—including standing ashtrays. Each design is critiqued, and the options are reduced to two. The remaining models are finished. The hardware is metal-plated, trademarks are designed and installed, and each model is photographed from every angle.

The scene shifts to the Nadir plant where test models are made. At the same time, handmade models are tested. "They are dropped, banged, hit, pushed, pulled, knocked around, frozen to subzero temperatures

Loewy and others await the presentation of the new Studebaker models at the company's South Bend, Indiana, headquarters. In his memoir *Never Leave Well Enough Alone*, Loewy describes in detail how to present a new design to clients, right down to placement of the ashtrays. (Raymond Loewy Archive, Courtesy Hagley Museum and Library)

and heated to tropical conditions. In other words, used and abused to the extreme." The next step in the process, after mass manufacture, is the product rollout. The designer describes a series of meetings with distributors:

> By the time refreshments have taken effect and tobacco smoke has reduced visibility . . . With a final blare of trumpets, the lights are turned out, spotlights hit the stage, the curtain rises, and the Nadir appears to the delirious audience in its smoke-veiled innocence. On its right is a gorgeous blonde in a gold lamé evening gown, on its left is a red-headed babe in a Bikini suit, with exceptionally long fluttering lashes. There is usually a moment of hushed silence, followed by a roar of appreciation at the beauty of the scene, mixed with many wolf calls and hiccups.

Sexism aside, Loewy likens the process to the birth of a child, but what he has delivered is a primer on design boutique operations. The second half of his lesson gives insight that other designers kept to themselves, how the firm can deliver a cutting-edge design that will please both the executives who commissioned the product and the purchasing public. Loewy, ever on the alert for a memorable catchphrase, dubbed the process MAYA—Most Advanced, Yet Acceptable.

Anyone who purchased a new Chrysler Airflow, Apple Newton, or anything designed by Phillipe Starck knows that great design does not always mean great sales. Loewy puts it this way: "There seems to be for each individual product (or service, or store, or package, etc.) a critical area at which the consumer's desire for novelty reaches what I might call the shock-zone. At that point the urge to buy reaches a plateau, and sometimes evolves into a resistance to buying. It is sort of a tug of war between attraction to the new and fear of the unfamiliar." Loewy emphasizes that the successful industrial designer should always know where the breaking point is for each commission taken.

To some extent, Loewy's use of the MAYA acronym is the equivalent of the modern-day executive self-help catchphrase—from *Most Advanced Yet Acceptable* to *Who Moved My Cheese*. Loewy establishes nine principles for MAYA:

1. A successful product that has been mass-produced over a long period of time will establish the norm for design for its product category.
2. Any new design that departs from the norm exposes the company to risk.
3. Risk to the company increases the gap between the norm and the advanced design. "For a big corporation, a little style goes a long way," he wrote.
4. Risk of a new design increases exponentially for smaller manufacturers and independent auto companies because they cannot effectively blanket the nation with a new product.
5. If the small company establishes a norm for a product, the larger company can often widen the gap between its current and coming models to set a design trend of its own. Conversely, large companies can choose to lessen the gap for a current design to reinforce its own look while disparaging the advanced design. "I wouldn't buy that stuff. It's too extreme. You won't like it." The larger manufacturer usually dominates sales through "sheer weight of manufacture."

6. The consumer is driven by two opposite factors when it comes to styling: (1) attraction to the new and (2) resistance to the unfamiliar. He quotes GM inventor Charles Kettering: "People are very open-minded about new things—so long as they are exactly like the old ones."
7. When resistance to the unfamiliar reaches its threshold and the marketplace sees resistance to buying, the product has reached the MAYA stage.
8. Loewy estimates a product has reached the MAYA stage when roughly 30 percent of buyers perceive the design as negative.
9. If the design is too radical, the consumer resists it whether it is a masterpiece or not. The company cannot force the public to accept a radical design. Loewy gives some constants in this example: teenagers are more receptive to radical ideas; two people with avant-garde taste will fall into more conservative buying habits if they marry; older age groups are influenced by the style opinions of the teenage group; wives are often the deciding factor at the time of purchase among married couples; and the MAYA stage varies according to geographic area, climate, season, income level, and other factors.

Loewy is clearly edging into satire with his readers, and the principles are vague enough so as to be applied to the design of practically anything, from a locomotive to a napkin. Still, at least he presents a philosophy. His advice is to push the design of an object beyond the familiar to the point where the design dazzles and surprises but retains enough of the original object's familiarity to breed contentment.

One Loewy contemporary, Henry Dreyfuss, had the bona fides to be a "great man" designer, but he subscribed to diligent research and dedicated his practice to defining how human beings use products and perceive design. In his book *Designing for People*, Dreyfuss's list of priorities for design is markedly shorter and nearly opposite in emphasis. Dreyfuss tells the reader that industrial design must have (1) utility and safety, (2) maintenance, (3) cost, (4) sales appeal, and (5) appearance. His view of the client is slightly cynical. "The business executive is looking for a man of vision who is not a visionary," he wrote. Dreyfuss used to enjoy illustrating his principles by drawing a hand and including each value on separate fingers.

In a prose style—featuring a witty mix of slang and jokey sentence construction mixed with more formal grammar—that remains consistent throughout the book and in letters from the Raymond Loewy Archive in the Hagley Museum and Library, Loewy tells the reader that all new designs involve some risk, and adds that many companies try to ensure sales by taking calculated risks. He categorizes several levels of risk. First, a large company can have decades of success by taking a minimum of calculated risk. This success will continue until a competitor sets a new design standard. Second, a smaller manufacturer can survive for years on minimal risk so long as the company closely mirrors the norm established by a larger competitor. But the company will not break the mold and forge ahead. "A case of pernicious sales anemia sets in, resulting in eventual extermination." Third, if pricing, quality, and engineering are in line, calculated risk is an open gate to improved business for a smaller manufacturer. Fourth, risk should never take the design beyond the MAYA stage except in the case of desperation, a tack he describes as "Commando styling," inspired by the medical term "Commando operation," where doctors remove huge amounts of cancerous bone or tissue in a last-ditch attempt to save the patient. Finally, there are ways to predict the MAYA level of a given product in a consumer market.

Just as in *Never Leave Well Enough Alone*, Loewy's correspondence in the Hagley archives features ornate locutions and grandiose, quirky word choice that suggest an affected "Americanism." The many reviews and news articles on Loewy written after the book's publication never mention a ghostwriter, although it's a good bet that Betty Reese had a hand in it. Although MAYA is pushed as Loewy's breakthrough mantra for his design philosophy, there are few instances where design historians can trace the evolution of a Loewy design from beginning to most advanced yet acceptable. The Loewy business archives from his most fruitful period, the 1930s and 1940s, were lost in an office fire. Loewy's insistence on keeping all sketches under his name also is problematic. Only the final approved sketch for each design was retained, bearing the Loewy signature, for company records. Most of the drawings showing a progression of design were, as Loewy designer Jay Doblin put it, consigned to "the waste paper basket." There are designs and paperwork from various clients in the two largest collections of Loewy's work—in the Library of Congress and the Hagley Museum and Library in Wilmington, Delaware, but there are hardly any complete files that reveal the genesis of a design from start to finish.

The most conspicuous surviving example of the MAYA philosophy that shows a progression of design is the Avanti sports car project. According to Paul Jodard, "Though it is difficult to place the drawings in a set order, it is easy to see how the design strove to get away from a traditional Detroit design—large radiator grilles, flared edges to panels, chrome highlights—toward a smoother, more European look. Certain features can be seen evolving, for example the flush sides and the inboard mounted headlamps, and the deliberate lift to the tail of the car. The gradual disappearance of the radiator grille can also be traced through the drawings and the models [for the Avanti]."

Returning to his lessons on client maintenance in *Never Leave Well Enough Alone*, Loewy, after detailing the tangible factors associated with running a design shop, takes an entire chapter to discuss the psychology of design. He describes the sensory aspects of design, such as why people continue to chew gum long after the flavor fades. Loewy theorizes that they keep chewing because they like having unlimited control over an inanimate object. He spends a paragraph or two discussing the way teenagers hold and sample their Coca-Cola bottles as compared to how an adult male holds a brandy snifter. The psychology of the Coca-Cola grip is not nearly as interesting as the psychology of Loewy's purposeful mentions of the bottle's design. Although Loewy often referenced the design in his public presentations, in *Never Leave Well Enough Alone* he doesn't lay any claim to the design as his own. Although he was more than happy to let listeners make the connection between designer and product, there are no bald-faced claims to creating the Coke bottle.

Loewy's "nudge-nudge, wink-wink" story of the Coke bottle "conception" can be traced to an issue of *Life* magazine from May 2, 1949, almost six months before the seminal cover story in the October 31, 1949, issue of *Time*. There doesn't seem to be the same level of original reporting in the *Life* article, although the writer dutifully recycles the designer's greatest hits—both professional and self-inventing. The structure of the article mirrors stories and sequences that would appear again two years later in *Never Leave Well Enough Alone*. All the bases are touched: the tailored army uniform, the Macy's window display fiasco, the Hupmobile account, and on and on. *Life* was the perfect magazine to establish this origin story because its circulation could reach one million issues per week, and the photo-dominant publication appealed to a wider consumer demographic. In addition, until about 1990, when an editor told a reporter to "pull the clips" (gather past articles) on a profile subject, *Time* and *Life* were almost always included in the search.

While the *Life* story implanted the Loewy legend to a national audience, the final few paragraphs give an intriguing glimpse into what the writer calls "the realm of Freudian fetishes." Loewy talks of the breast-like shape of jelly molds and the "warmth and fleshiness of plastics." At the end of the article, the writer tells readers that Loewy "broods a great deal about the callipygian Coke bottle."

Things get considerably more Freudian deeper into the final paragraph, because "callipygian" refers to the classical statue of Aphrodite Callipygos, which is described by Stephen Bayley as "the starting point for any cultural history of the eroticized bottom." Loewy describes the bottle's shape as "female" and would continually refer to the Coca-Cola curve throughout his career and in his designs—most notably in the Avanti. The auto industry used callipygian curves in major designs, including the Chevrolet Camaro and Ford Mustang. Interestingly, *Life* writer John Kobler gives full credit in the article for the bottle's design to Alexander Samuelsson. The sexualization of the female shape did not begin or end with Loewy. Abstracted feminine shapes appear in bottled products such as Deskey's Joy dish soap, Wesson vegetable oil, and a variety of perfume bottles. Over time, as Loewy made more speeches referencing the design and more reporters and critics referenced the *Life* article, Samuelsson's name faded out of the "designer of the Coke bottle" legend.

Even the Raymond Loewy website raymondloewy.com, overseen by his estate, lays claim to his authorship of the design. It's generally accepted that Loewy and his design team created a new "slenderized" look for the bottle, but historians object to claims that he "created the Coke bottle." His work in replacing the logo embossed in glass with a stark white lettered logo certainly modernized a classic look. Still, that is a far cry from inventing one of the most instantly recognizable packages ever conceived.

The all-red, sleek fountain and vending units he did design were nationwide symbols for the soda and today are nearly as iconic as any bottle designs. Loewy gave the fountain dispensers a soft, streamlined shape that recalled the sleek designs of Brooks Stevens's outboard motors. Even more ubiquitous was Loewy's Coca-Cola "coffin" bottle-dispensing vending machine, which was seen in gas stations, convenience stores, and arcades across the country. In large part a horizontal version of the rounded Sears Coldspot refrigerator, the Coke dispenser was one of the easiest vending machines to use. Other designs that made it into production were new cans, labels, and cartons.

Loewy's sure touch with putting together a campaign extended to creating competitive and smoothly functioning teams. Betty Reese revealed Loewy's employee management style in an article in a 1986 *ID Magazine* retrospective after the designer's death: "He was also a master manipulator of his staff. He could see dissatisfaction, he could smell disloyalty. He would take whoever was unhappy aside and gently say, 'Now, Mike, you wouldn't be thinking . . .' Or he would lean over their shoulder at the drawing board and say, 'Oh, what a beeeautiful drawing' and whoever it was would just melt." The innate courtliness of Loewy's

Raymond Loewy's work for Coca-Cola focused mainly on packaging and creating the sleek, red dispensers and vending machines that still today are instantly recognizable as Coke products. Loewy also designed a "slenderized" version of the Coke bottle, but contrary to current claims, he did not design the original. The claim arose from a 1949 *Life* magazine article that praised the bottle's eroticized "Callipygian curve." (Raymond Loewy™/® by CMG Worldwide, Inc. / www.RaymondLoewy.com)

personality was a critical part of his supervisory skill set. Reese again: "Loewy adored draftsmen. He would make a little sketch that was really quite primitive but which expressed his thought, and then his designers, who were terribly talented, would translate it into a beautiful drawing. The office would sometimes put on shows in which they'd spoof Loewy's inability to draw or even his superficiality, but he was always the one to laugh loudest: he appreciated the fact they were on to him."

The function of a design studio does not require a leader with self-effacement or exceptional draftsmanship. Rather, as Loewy and Harley Earl proved, the most important function of a design shop leader was as a sounding board and editor. A handful of the other design pioneers were not brilliant draftsmen. Dreyfuss probably had the most innate artistic talent, producing excellent sketches, renderings, and proposals, particularly in the first years of his company's existence. Dreyfuss developed a shorthand sketching style and passed his ideas along for his renderers to conceive. One skill Dreyfuss cultivated was the ability to sketch upside down for clients sitting across the table (it was an impressive presentation skill, even to the most jaded client). Paul Williams, a now-celebrated African American California-based architect, learned upside-down drawing to spare his clients from having to sit beside him.

Despite Loewy's penchant for credit grabs, most of his staff realized that working for Loewy and Associates meant the guarantee of working on a wide range of commissions. In addition, when they found a design niche that would deserve their full-time attention, Loewy gave them enough space to establish a reputation while laboring under a large corporate umbrella that could absorb a missed commission or client desertion without folding. "His people came to think of themselves as the design elite in their various specialties: the top designer of products, packages, ships, planes, trains, was without question a Loewy man, and this formidable arrogance permeated the Loewy organization," wrote Betty Reese.

Loewy ends his lessons in agency management by instructing would-be designers how to create an automobile design studio. In the only extended reference to Studebaker in the memoir, Loewy informs that the designer's job is to (1) establish a long-range design philosophy for the company, (2) develop a practical design for all models within that philosophy, and (3) maintain an atmosphere of creativity. The lessons continue as Loewy advises designers to work on several problems at once (to maintain a creative edge), work flexible hours, and get out of the office or plant to travel. "By giving the men a chance to escape their regular surroundings, to work at a different pace, with different associates, their outlooks are refreshed. The world seems to bring fresh ideas and sustain their creative talents at peak level," he writes. This represents a marked change in philosophy from the Big Three automotive design studios, where designers were often forbidden to leave the office or even visit other studios within the same company.

Loewy leaves nothing out in his advice on running a design firm, not even sales calls and business trips. "Industrial design, as far as I'm concerned is 25 percent inspiration and 75 percent transportation," he wrote in a chapter that outlines a business trip in exhaustive—and, frankly, boring—detail. He leaves out nothing, not the crying baby in the airplane, not the trip to a South Bend burlesque joint, not the details over every meal eaten on the journey. The intent was to entertain the reader, who is imaginatively called "Mr. Reader," but the lesson delivered is that to be successful an industrial designer, particularly one who owns a substantial firm, one *should* spend little time designing anything.

Travel, meetings, and sales calls are the lot of the designer-owner, who can do little to change it other than designing more efficient ways to meet obligations. Loewy teaches that sometimes the reason a company becomes successful is less about the skill of the employees and more about the willingness of the boss to entertain and service clients. To be successful, it's also important to practice the art of entertaining before the art of design.

A designer should also be somewhat enigmatic and mysterious. If his memoir philosophy was "never leave well enough alone," then his management philosophy might well have been "never let your employees see too much of you." "No one had much contact with the man, even then," wrote Jay Doblin, who started working at Loewy and Associates as an office boy in 1939. "He was rarely in the office—he was always somewhere else in the world. His schedule was to go to one of his custom-built houses for three to four months in the summer, come back, and spend three weeks in the New York office to check on what was going on and then fly off to spend the winter in Mexico or Palm Springs or St. Tropez, where he anchored his 85-foot yacht."

The firm was a "benign dictatorship," in which all employees were members regardless of title. Loewy owned the entire company. No stock was issued, nor was there any profit sharing except for full partners. This business model held true for three of the Big Four designers: Henry Dreyfuss, Walter Teague, Norman Bel Geddes, and Loewy. Only Dreyfuss, who preferred to personally supervise every account, limited the amount of new business that he or associates brought in. "From the outset I determined to keep our staff small and compact, so we might render a personal service to our clients," he said in a speech before the Harvard Business School.

By maintaining a close-knit staff of highly specialized artists and account managers, Loewy was able to call on the acute perceptions of designers with an innate sense of what was "most advanced." By pushing the envelope with designs by four or five competing designers, Loewy's staff could quickly reach the "most acceptable" design, thereby avoiding major mishaps. Additionally, the editing skills of Loewy and partner designers Barney Barnhart and Bill Snaith gave the company a collective strength that certainly outstripped much of the competition.

Loewy himself becomes the most advanced image of the "great man" industrial designer, while his careful burnishing of his reputation never went beyond acceptable bounds.

> Loewy's insistence on the importance and relevance of industrial design somehow justifies his relentless self-advancement, if not his slightly opaque attitude to clear facts. . . . Loewy had been in the advance guard, had set out the ground rules for the profession, had made the dazzling deals and won the accolades. He intended to go on being there as long as he could, as well. At the age of 70 he took a high-speed driving course with the American racing driver Carroll Shelby: seven years later he is roaring over California beaches in a dune buggy! The last years of his life sound almost like a fight for youth and longevity. He had always been up in front, and most advanced he was going to stay.

It was during this period, from the late 1960s onward, that Loewy laid the seeds for his eventual reputational renaissance. He began doing more "elder statesman" interviews, and as the last surviving industrial design pioneer, he had the playing field to himself. Laurence Loewy recalled her father's lust for life in an anecdote published on raymondloewy.com. Her story opens as her mother, Viola Erickson, gallivants through Palm Springs shops with visitor Truman Capote. Laurence is pulled in to copilot her father's dune buggy. "Dad would have made the 10 best-dressed list with his custom cowboy boots, pants, shirt and hat. I on the other hand, made the mistake of wearing shorts and sneakers." The dune buggy sported a long flagpole to alert unsuspecting onlookers of the approaching dervish. "Dad was fearless. He attacked the steepest dunes with pedal to the metal ferocity. The buggy was powered by a high-performance VW engine quite able to produce enough speed for the terrain and send the local wildlife dashing for cover." Laurence recalled that Loewy's manic driving could be less than entertaining. "On numerous occasions the rear tires failed to grab enough traction thereby sending Dad and I sliding backwards down a high dune. It was at those times Dad would send me off to gather anything that would enable the buggy to push off and clear the dune's crest. I'd walk, dripping with sweat, with my sneakers filled with sand and load up on driftwood, rocks and painful cactus quills. Gripping the steering wheel, Dad would say 'The race must go on!' "

Loewy's every decision, every story in the media, every article of clothing, every personal vehicle, and every home functioned as an advancement of the ultimate lifestyle. His possessions and work brought in business and advertised Loewy's taste. Most of the profits from Loewy's firm fueled his lavish lifestyle, which went a long way toward selling the services of the company. Once his services began to be recognized by the consumer public, Loewy set out to increase the value of publicity. On many levels, Loewy understood that manufacturers were not buying new designs from him—they were buying the promise that a new product, filtered through Loewy's personal taste, would succeed in the marketplace. Shortly after expanding his office following World War II, Loewy realized that the most important product he sold was his image as a man, for lack of a better description, "of wealth and taste."

The *Time* cover story was a brilliantly presented idea of an American "design for living." Postwar America would come to devour these articles, reading of the lifestyles of the famous, from Loewy to Diana Vreeland to Martha Stewart. One irony in the cover photo is that the treatment, which features a headshot (framed in red) superimposed on backgrounds related to the subject's profession, was conceived by Henry Dreyfuss when he redesigned *Time-Life* publications for Henry

Luce. Placed in the *Business and Finance* section, the story, headlined "Up from the Egg" (a reference to Loewy's frequent claim that the egg was the world's most perfectly designed product), is placed under the subtitle "Modern Living."

The article opens with a litany of Loewy's furnishings and possessions. In the first paragraph, a flip of an electric switch splashes indirect lighting "over walls made of egg-crate fiber" and an inventory of exotic possessions: a Tahitian drum, Congo ceremonial sword, Chinese helmet, Moroccan fly-switch, Senegalese war hatchet, and "grotesque Zulu masks." When the designer awoke, presumably with the *Time* reporter observing from a chair, Loewy entered his "black, beige and bronze bathroom, with its motif of Nubian slaves." He dons an "expensively tailored grey suit" that features replaceable inch-and-a-half cuffs. Following the designer, the writer describes the living room as "glittering with thousands of flecks of gold-colored plastic thread woven in chairs, sofa and carpet." The photograph illustrating the opening page could be taken from a Nick and Nora Charles film. Loewy is dressed in a satin smoking jacket, ascot, and slippers, complemented by starkly patterned, bold socks. Loewy's wife Viola (the couple had been married less than a year), dressed in a flowing, floor-length robe, takes coffee from Karl Huzala, the couple's butler. Huzala serves in front of a wall so crowded with art that there is barely room for an electrical outlet. No postwar reader could take in this level of detail without imagining a world where he too could own the latest products, marry a gorgeous ex-model, and live in an apartment decorated like the Louvre.

There was plenty of inspiration for readers to slip into fantasies of consumption. Across the country, more than one million new homes were constructed each year, all of which had to be outfitted with furniture, appliances, housewares, and cars in the garage or driveway. Those cars, of course, were rolled out in new versions every year. That *Time* chose Loewy to epitomize this new movement, where industrial designers were creating products to fuel the portrait of success, was a testament not only to Betty Reese's public relations acumen but also to the esteem that designers had earned.

Three of the Big Four, Loewy, Teague, and Dreyfuss, all built their businesses to the height of accomplishment during the immediate postwar period. Even relatively unknown designers created agencies with dozens of employees and branch offices. Donald Deskey, for example, who had been slowly centering his design business on packaging and product design, was able to triple the size of his company. He opened branch offices in Brussels, Copenhagen, London, and Stockholm and served clients such as British Leyland, DuPont, General Electric, and Westinghouse. In the 1940s, he employed six; by 1960, he was overseeing a hundred employees. His philosophy of business building was no different than Loewy's. In a lecture at the Massachusetts Institute of Technology, he said an industrial designer "was no living embodiment of Leonardo Da Vinci, but a hard-working organization equipped to render a needed service to industry."

The *Time* article details Loewy's day at work, taking meeting after meeting, stopping to snip out a package design using scissors and colored paper. He skips lunch and opts for "a rubdown and massage at the New York Athletic Club." The writer sensed that Loewy orchestrated his own image. According to the article, Loewy talks in a subdued voice that is at the same time apologetic and compelling. "His face is described as reposed, gentle, sad and as inscrutable as a Monte Carlo croupier. Obsessively shy, he is always 'Mr. Loewy' even to his closest associates. Even to those who know him well he is something of an enigma. Said one long-time acquaintance: 'After all these years, I'm not even sure that I like him!' Everything he does calls attention, with skilled showmanship, to his work, so that observers at times get the strange feeling that he too is a design—by Loewy of course."

Loewy was kept busy no doubt because he was scrambling to keep up cash flow to afford the upkeep on no fewer than five homes. The homes, like Loewy's personal style, were just as much a marketing statement as domicile. Unlike Benjamin Sonnenberg Sr., who focused all of his image-making in New York City, Loewy deemphasized the apartment in New York. Each home seemed calculatedly decorated to push the edge of contemporary taste. He owned a stark, austere modern home in Palm Springs, a home in Mexico, three homes in France, and a Manhattan penthouse. The *Time* article describes the New York apartment's décor like an auction list at Sotheby's, culminating in an inventory of Loewy's "art wall," which includes a Joan Miro, a Picasso, a Matisse, and a Raoul Dufy painting mounted on a hinge that swung open to reveal a hidden television set. This litany of luxury is a template for the "lifestyle" reporting so prevalent in the last part of the twentieth century in such "shelter magazines" as *Architectural Digest* and many others. Loewy, whose self-awareness was nearly as acute as his ego, wrote, "I'm sure my lifestyle was an easy target for other designers. A good life has been as important to me as my work."

The good life Loewy most prized was represented by the Palm Springs house, designed in 1948 by architect Albert Frey. He devotes one of the final chapters in *Never Leave Well Enough Alone* to his house, dedicating nearly every word to how the building reflects his taste. He described picking the site: "I felt very happy, and I walked a great deal through the desert at the foot of Mount Jacinto. On one of those treks, I found an interesting spot on the side of a low hill. It was a maze of granite boulders, some of gigantic dimensions, all pale gray in color, part of a prehistoric glacier. I thought it would be a perfect site for a small desert retreat."

Frey would remember that Loewy entered his office in 1946 with the site already chosen. As Frey worked on the house design, each drawing delivered to his client would be returned by a Loewy factotum, detailing changes. "Loewy used to stay at the [Palm Springs] Racquet Club," said Frey, who also designed the tennis club. Loewy's chosen lot was a second choice; he would have preferred to build in a development known as Little Tuscany, but that area had a homeowner's association that forbade anything but "Italian" design. He bought two acres on a lower sloping lot surrounded by citrus, banana, and palm trees and cactus. "A lot of the rocks had been pushed onto the property when they were building the roads. So we left most of them. He wanted to use them and they made the project more interesting," Frey recalled.

The Loewy Palm Springs house was small (the designer was divorced at the time), built for $60,000. An article in the *New York Times* described it as a house "for a solitary man." "He didn't even have a phone" Frey said in the article. The small desert retreat had a focal point, the free-form swimming pool, which was marked off by four huge boulders that were partially immersed in the 80-degree water. But unlike most of the swimming pools in Southern California, the pool did not end in the backyard. It reached past the 30-foot sliding glass doors and ended in the middle of the living room, a few feet from the front door.

Guests to the Loewy home would often fall in. At the housewarming party, actor William Powell, noted for his urbane role as Nick Charles in the *Thin Man* comedies, toppled fully clothed into the pool. Singer Tony Martin, seeking to save Powell some embarrassment, leaped in as well, closely followed by the host.

The famous "inside-outside" swimming pool at Loewy's Palm Springs home. Visitors could swim into the living room, and many party guests ended up in the pool, often fully dressed. (Raymond Loewy™/® by CMG Worldwide, Inc. / www.RaymondLoewy .com)

"It was an interesting collaboration with Loewy because he had wonderfully inventive ideas," Frey said. Among Loewy's interesting ideas: using pecky cypress paneling (a wood noted for its "wormhole" look) and redwood for a trellis that ringed the pool area and framed the desert vistas from the pool and other rooms. A pane of thick, corrugated glass was the only barrier to the unobstructed views of the backyard. Frey also used pecky cypress in a grid framing the master bedroom and bathroom windows, and built glass block shelves so that Loewy's collection of ceramics and glass could be displayed at night. When he remarried, Loewy added a small work study and created a simple large bedroom addition to the original house. The home, dubbed *Tierra Caliente*, a name Loewy also would give to another residence he owned in Mexico, was his base of operations in the winter months and a semi-permanent home (with *La Cense*) once he retired from a full-time role in his firm.

In his memoir, Loewy lamented that the life of a designer was so demanding that he needed to decompress with periods of absolute rest. Without a trace of irony, he suggests taking several weeks in the desert and spending several months in Europe. His choice of Palm Springs as a desert retreat was an aesthetic choice and a shrewd business invest-ment. "What I liked most was the subtlety of the coloring," he writes. "The scents and the sounds; subdued, restrained, as if nature had blended everything into a delightful understatement. It is a symphony of delicate grays, beiges, horizon blues and lavenders; the textures, whether harsh or delicate, are invariably exquisite."

Somewhat less exquisitely, Loewy the memoirist gives the impres-sion that he designed the home, shortchanging Frey, whose nationally known work was celebrated well before Loewy hired him. Loewy re-counts returning from a Racquet Club dinner (with Hoagy Carmichael) and making a rough sketch of a home. As always, Loewy claims a rough sketch contains the creative edge of inspiration that is carried out by a lesser light. "I called on a local architect of whom I had heard," he writes. Frey acknowledged Loewy's input, but the designer's account makes it seem as though Frey merely embellished Loewy's overall design.

The Palm Springs setting was a perfect backdrop for business and presentation of Loewy's ultramodern taste. Loewy's first contact with President John F. Kennedy came in Palm Springs when the president flew in to visit in 1963. Their meeting inspired several commissions that would rank among Loewy's most famous late-career designs.

The designer's typical routine in Palm Springs is detailed in *Never Leave Well Enough Alone.* He starts the day with a soak in the "Indian bath," a natural hot spring. After a parboiling, he returns home for a nap and a swim, all before 8:30 a.m. He spends the rest of the day making business calls and working on designs. The evening is spent with friends. "Jack Benny likes to spend a quiet hour with Viola and me, and we all relax and feel happy. Or it might be a movie director, or a starlet, or a star. William Powell, our neighbor, is often with us, with Mousie, his delightful little blonde wife." Loewy described the feeling of living this way: "When the house is kept in darkness, save for a log fire and candles, the sight is sheer beauty. A small fountain adds its frail tone to the silence of this oasis. In the distance, we hear the coyotes. Viola is near me. R.L. is happy."

Loewy and his wife spent most summers in France, part in their estate home *La Cense* and the rest of the summer in "the fisherman's village of Saint-Tropez." The Loewys bought the villa, called *Uriane*, in 1930. The larger estate *La Cense*, bought in 1933 at the height of Loewy's initial success, was built by King Henry IV in the sixteenth century. The rambling manse, the homiest of Loewy's many residences, was built by the French monarch for a mistress, Gabrielle d'Estree. The Renaissance manor featured a large interior courtyard.

The photographs of the house reveal a space that is lived in, not staged. He built a "café" in what he described as a refurbished henhouse. It featured a massive music box, an Edison phonograph with a stack of wax cylinder songs, and a crank-operated telephone purchased in 1910. The space even had an amusement park "shooting gallery" where guests could shoot at targets. That is not to say that Loewy tossed furnishings willy-nilly throughout the rooms. On the walls were works by Pablo Picasso, Max Ernst, and Marc Chagall.

"We've always lived life to the hilt," Loewy told the *New York Times*. "Money's only value was what we could do with it." Still, the designer was not above creating a marketing effect with his French homes. "We are surrounded by our friends, the Duke of Brissac, the Count of Pourtales and Denis Baudoin," wrote Loewy in *Industrial Design*. The house had a large interior court surrounded and complemented by a pond bordered by weeping willows. Loewy maintained a design office in the attic.

Uriane, in Saint-Tropez, was "a summer home situated at the water's edge on the side of a low cliff." The house was built on columns directly over the ocean and had a small dock jutting into the Mediterranean Sea, where Loewy moored a succession of cabin cruisers. Loewy designed the interior using Japanese motifs as design inspiration. He bought most of the decorative materials on a trip he made to Japan at the behest of Stuart Symington. He brought in black lacquer panels, lighting fixtures made of white rice paper, bamboo vases, and silk scrolls. The same Japanese motifs as the Saint-Tropez villa served as the décor for the designer's attic workspaces in *La Cense*. His studio overlooked the Mediterranean. He shipped over to France an American speedboat, which he named *Ayrel* after his model airplane design, for use on the Riviera. The robust athleticism that Loewy cultivated all his life is detailed as he tells of taking out his boat *Loraymo II* (he used the name *Loraymo*, based on his telex address, over and over for cars and boats) and deep-sea diving for hours on end. The tale of how he designed his diving experience first appears in *Life* in 1949. Loewy complains that the air hoses transmitted a nauseating oil smell as they baked in the sun on board the dive boat. His solution was to add a few drops of Chanel No. 5 or Stradivari to the oil to dissipate the odor. (This story would also become a staple of Loewy's marketing portfolio.)

Loewy's Mexican residence, in Tetelpan, a sleepy village outside Mexico City, also carried the name *Tierra Caliente*. "I erected a house I actually designed in New York," he wrote in *Industrial Design*. The house, which featured 18-foot ceilings, had pink marble floors on the terrace and throughout the house. The living room had alternating strips of black-and-white marble. The interior was so striking that *House and Garden* published a cover story on the home.

"Thus we live a life which, to us, is ideal. It is a blend of everything that makes life interesting and eventful," writes Loewy of his life with Viola. "America gives me the opportunity to be creative and imaginative. Europe—and France in particular—brings relaxation and perspective. This slowing down is imperative in order to maintain a balanced outlook. It also gives me a chance to appreciate America more keenly." Every residence, save perhaps *La Cense*, the French villa that seems more like a home than a showroom, was used to hone the designer's public image or to impress prospective clients. The intensive fees that Loewy raked in year after year fueled this extravagant lifestyle, but who is to say that the marketing of Loewy's personal style did not fuel the business itself?

Tierra Caliente, the famed Palm Springs, California, home designed for Raymond Loewy by architect Albert Frey, was a visual calling card for Loewy's lifestyle. He used the home as a bachelor getaway when it was first built, and then turned the modernist house into an entertainment center for clients, celebrities, and friends. (Raymond Loewy™/® by CMG Worldwide, Inc./www.RaymondLoewy.com)

12

The Sales Curve Wanes

A t the end of Loewy and Associates' contract with Studebaker, in the late 1950s, the landscape of industrial design was changing, particularly for the independent shops. Most large manufacturing firms realized that they needed well-designed products to survive in the marketplace, and that they could hire in-house designers to create entire lines of products, eliminating travel fees, arguments with outside designers, and problems with consistency throughout a differentiated product line.

Loewy, ever the businessman, realized this trend early, although he still publicly railed against specialized designers, well before the Studebaker contract ended, and he began to take on more challenging assignments in nonmanufacturing fields to offset the inevitable falloff of clients. It would be unlikely he could ever sublimate his personality to single client or industry. Instead, he sought more fertile fields in markets that might be behind the mass-market design curve. His client list later went international when he reopened his London office in 1948 and established his French company, Compagnie de l'Esthétique Industrielle (CEI) in 1952, later signing Shell Oil and British Petroleum to long-term contracts. Europe and other markets lagged behind the United States in initiating marketing, business, and design trends.

Part of the company's decision to branch out took place during World War II. Loewy and some of his colleagues found lucrative contacts for future assignments on governmental projects. Loewy designed several commissions for the new intelligence agency, the Office of Special

Services. In one of the few multiple-colleague design collaborations of the war (the most memorable being the Jeep design by several car companies), Loewy, Henry Dreyfuss, and Walter Teague worked together to create the strategy room for the US Joint Chiefs of Staff. In addition to presentation facilities, the room featured a 12-foot-by-24-foot world map painted on a curved surface. The room also featured four rotating globes designed by Dreyfuss—one each for Josef Stalin, Winston Churchill, Franklin Roosevelt, and the Joint Chiefs. During World War II, these designers and others found commissions from every branch of the armed forces and continued to bring in new work from civilian companies whose products had to adapt to wartime shortages and conditions. By the end of the war, designers had found a lucrative new market in federal commissions, and design companies realized that design jobs could be extended beyond commercial products.

One such commission, although relatively minor in the Loewy firm's body of work, inspired Loewy to coin his oft-quoted claim that he designed products from "lipsticks to locomotives." (Loewy even used it as a subtitle for his book *Never Leave Well Enough Alone*.) Cosmetics companies during the war were prevented from manufacturing metal cylinders for lipstick. A small company, the House of Westmore, asked Loewy to design an alternative. The design staff produced a cardboard tube that used surface coatings to resemble metal. "In creating the container, I like to think I did something for American men as well!" Loewy wrote. The shortage of most metals, used for the manufacture of ammunition and war materiel, led to the development and increasing use of plastics and glass, trends that would continue to dominate packaging after the war. The 1943 lipstick design remained a topic in many of Loewy's lectures and publications until his retirement.

The decline in manufacturer commissions did not happen overnight. Like most business trends, it took place gradually, and then, at the end, suddenly. By the end of the era of the huge multiservice design firms, design services had not disappeared; instead, they had morphed into boutique practices that offered one or two areas of expertise. One of the Loewy firm's most acclaimed designs in the immediate postwar era came from a company noted for internally designed products—a vacuum cleaner for the Singer Sewing Machine Company. By contrast, before the war, Dreyfuss created what would become the benchmark of streamlined home appliance design in the rounded, swooping design of a Hoover upright vacuum. Loewy's design for Singer eschewed pure streamlining for a more blockish modern design influenced by the emerging International Style of architecture. Loewy's designers created a visionary design that was "about the height of a pack of cigarettes" and that could easily drop its handle to reach under the lowest of sofas. The easy-drop handle also made it simple to store the vacuum on a hook in a pantry or closet. The machine was featured in the 1948 *Annual Design Review*, but it turned out to be a failure in the marketplace. Singer's reluctance to spend on advertising and marketing helped usher the design into oblivion.

The Singer vacuum was a rare Loewy failure in a booming economy where housewives were buying any cleaning machine that could be plugged in. The postwar market, while robust, complicated consumers' lives through the many product choices made available to families. According to *Domestic Revolutions*, by Steven Mintz, the 1950s housewife spent much more time on housework than her mother or grandmother. According to a 1945 survey, a farm wife spent 60 hours per week on housework, while an urban housewife spent 80 hours per week on household chores. Although every advertisement for household appliances touted vast time savings for the fortunate housewife, the appliances actually increased the homemaker's commitment to housework.

As companies sprinted to take advantage of the bull market for design work, the bigger design companies realized it was advantageous to make sure that industrial design was recognized as a bona fide profession. Together with Dreyfuss and Teague, Loewy helped organize the Society of Industrial Designers (SID). "There were giants around the table," said Dreyfuss of the first meeting. The professional society was focused only around the independent designers who were already the giants of the profession. Any designer could join—in theory. The society was often referred to as an exclusive club rather than a professional organization, at least in the early years. To join the society, a designer had to submit a portfolio of work to the board. The board had to unanimously elect the candidate to the board. Brooks Stevens recalled, "I remember vividly the feeling of pride I had in being invited to join the gods on Mount Olympus at the first meeting in New York."

SID arose out of the meteoric rise of design offices during and after the war. A series of high-profile articles on Teague, Loewy, and others eventually caught the attention of the New York Tax Service. The state tax agency sought to tax design studios under the state's unincorporated business statute. The designers claimed they were exempt from the higher tax rate because they provided professional services, much like an architect or attorney. The designers chose Teague to put forward a test case in 1943. Teague's attorney argued that industrial design was just as much a profession as law or medicine, citing Teague's 1949 book *Design This Day* as proof of its legitimacy. The specific language of the tax case stated that tax exemption was permissible if "eighty percent of the gross income is derived from the personal services actually rendered by the individual." The decision essentially legitimized the "great man" model for design firms. The firms led by Loewy, Dreyfuss, and Teague epitomize this model, which called for a charismatic lead designer who solicited business, oversaw commissions, and sold the final product to the client.

The tax decision led directly to the establishment of the Society of Industrial Designers. The American Designers Institute, which was incorporated as an association of designers working in the furniture industry, already existed, but its dependence on membership from the craft industries did not project the exclusivity that the industrial designers desired. The rules for membership restricted the association to those who were "successfully engaged in industrial design" or academics teaching industrial design. A designer was defined as one "who has successfully designed a diversity of products for machine and mass production." The cleverly worded clause excluded those working in single industries.

The wording of the membership rules caused immediate friction between the two groups. At the society's first meeting, celebrated furniture designers George Nelson and Charles Eames were not offered memberships. Loewy was still decrying in-house designers in his 1979 *Industrial Design*, saying, "Captives! Following the orders of marketing managers makes them social servants."

In much the same way that American car executives resisted smaller cars and sleeker designs and railroad companies refused to recognize the inroads of other types of transportation, the leaders of SID felt that in-house designers and designers working in a single discipline were somehow a lesser form of artist. Loewy said succinctly at a meeting of the society in 1959, "Designers should not specialize," adding that designers who did just one product were not industrial designers. Loewy and others believed that designers working within a corporate system would eventually be absorbed by it, thus losing the outside evaluative eye that is valuable to the company. "If designers get reabsorbed, digested, ingested, mutated or reoriented by the action on them of non-designing forces of executive enzymes there will be no industrial design profession," Loewy said.

Despite Loewy's most dire warnings, the ranks of corporate designers grew, apparently fed by executive enzymes and then by increased marketplace demand. By the beginning of the 1950s, the heyday of the full-service design boutique, epitomized by both the Loewy and Dreyfuss operations, was waning. Manufacturers were looking to market research to formulate strategy and wanted to forego the advice of an omnipotent expert who relied on a personal vision to create the product. As the market slowly changed, more independent designers began to drop areas of specialization to concentrate on the core strengths of their firms. Dreyfuss, whose company maintained a wide range of special services well into the 1960s, eventually turned away from solely turn-key assignments to concentrate on ergonomics. In addition, the look of the product—the packaging—save for certain industries such as automobiles and consumer appliances, became just as important as the product. In the postwar era, Donald Deskey, the creator of Art Deco and skyscraper furniture in his early years as a designer and of classic 1930s interiors, including his masterpiece Radio City Music Hall, became almost exclusively a package designer for Procter & Gamble, creating packages for Crest toothpaste, Joy dish soap, and Tide detergent. Loewy faced these same pressures, but Loewy and Associates, which already had essentially divided itself into areas of specialization, such as Bill Snaith's retail and interior design division and the packaging division, was able to ramp up those areas to pick up the declining assignments in other areas.

Packaging spawned not only a new arena of business, but also specialist designers who created nothing but packaging. Package design hardly emerged full-blown from the minds of the streamliners of the 1930s. The instantly recognizable yellow-and-black packaging of Kodak products was the brainchild of Walter Teague, but the red-and-white Campbell's soup can and the Coca-Cola bottle all qualify as design that is so familiar that consumers automatically associate entire product lines with a single package image.

Packaging as a sales tool came into use in the 1920s, concurrent with the rise of advertising on radio and in four-color ads in national (and popular) magazines. Combined with an expanding system of roads and railways, these factors brought together conditions to create a large market for products nationwide. Previously, the package served as protection for the product as it made its way from plant to store. But this primary function was superseded by the need to use the package to market and advertise the product within. One of the first national designers to specialize in package design was Ben Nash, who operated a New York City-based company that employed more than thirty designers. One of his earliest commissions was a product line redesign of Armour meat products (which would be redone in 1944 by Raymond Loewy).

The first modern packages were metal tins—often containing medicines, elixirs, and salves—with decorated labels, an innovation that historian Thomas Hine traces to seventeenth-century London. In America, the concept of packaging products took hold around 1875. Railroads allowed goods to be sold widely and cheaply, and thus manufacturers needed consumers to remember their regional or national products. At the same time, the migration from family farms to urban centers took consumers away from the tight-knit influence of family and exposed them to newspapers, magazines, catalogs, and other media that carried advertising. The first real breakthrough in packaging was the paperboard folded box, which ensured that whatever product was contained inside would be less susceptible to spillage and damage. Boxes also could be easily printed and stacked in store displays.

Loewy's first major package commission came well before the design professions emphasized the packaging niche. After the 1939 World's Fair, the chairman of the American Tobacco Company, George Washington Hill, an eccentric man fond of wearing (in business meetings) a wrinkled hat festooned with fishing lures and hand-tied flies, came to the designer's office. "I would not call him a rounded man," said adman Albert Lasker of the Lord & Thomas agency of George Hill. "His only purpose in life was to wake up, to eat, to sleep, so that he'd have the strength to sell more Lucky Strikes." Hill kept Lucky Strike packages taped to the rear window of his Rolls Royce, and he doted on two dachshunds, named Lucky and Strike.

The cigarette assignment was yet another commission materialized out of cocktail party banter. As Loewy tells it, Hill asked Loewy about a remark the designer had made to Lasker deriding the package design of Lucky Strike cigarettes. Hill and Loewy traded repartee and compared their Cartier suspenders before Hill asked, "What about that package? Do you really believe you could improve on it?" Hill had deep-rooted opinions—on everything. He once told Raymond Rubicam, the lead partner of Young and Rubicam Advertising, that he entertained his grandchildren by showing them his pocket watch, his pocket knife, and his dentures. "The public's reaction to entertainment and advertising is no different. You just don't understand the advertising business," he told the adman. Rubicam finally resigned the $3 million account after Hill requested he replace the entire creative team for the fourth time. Back at the office, Rubicam's employees reportedly danced in the aisles.

Back at the Loewy offices around 1940, the two men wagered $50,000 on the project. Hill accepted the "bet," offering Loewy a $20,000 retainer (about $352,500 today) and, "if we use the result, $30,000 more." Hill asked when the design would be ready, with Loewy replying, "Oh, I don't know, some nice spring morning I will feel like designing the Lucky package and you'll have it in a matter of hours." Loewy described challenging Hill to work with colored paper and scissors in an attempt to improve Loewy's design. This "challenge scenario" became a prominent motif in later Loewy speeches and recollections. Although it is difficult to believe that the president of a multimillion-dollar corporation would spend his business time essentially playing with school supplies, when it came to the designer-client relationship—as it applied to American Tobacco at any rate—Loewy was willing to sink to any level to make the sale.

Lucky Strike held a fascination in American popular culture, particularly with the modernist movement. Stuart Davis, perhaps the greatest pop artist and certainly one of the first major artists to incorporate popular imagery into his work, in 1921 focused an entire painting around the popular and simply modern design of the all-green, pre-Loewy Lucky Strike package. Loewy's redesign of the cigarette package is a story of simplification. The original package was dark green, complemented on one side by the cigarettes' trademark target logo with the brand name printed inside. On the back of the package was another target, this time filled with minuscule text. In the redesign, the text-heavy back side was eliminated, switching all of the text to the sides of the package. The green ink, which was expensive and exuded an overtly chemical odor, was eliminated in favor of a white background. Finally, Loewy repeated the red target on the back of the package, thus ensuring that the brand name was visible regardless of how the package was laid on a surface. "Putting

Loewy's most celebrated package design took a cigarette pack dominated by a sickly green-and-red target and created a clean, hygienic design that switched Lucky Strike's background color from green to white. The design, which was used on both sides of the package, was easily recognizable to customers (and sales clerks). (Raymond Loewy™/® by CMG Worldwide, Inc. / www.RaymondLoewy.com)

the logo on both sides of the box was Loewy's genius," writes Philippe Tretiak. "The simplest answer is best."

"The white background made the brand symbol more conspicuous," wrote design historian Adrian Forty. It automatically denotes freshness of content and immaculate manufacturing. "Hygiene, cleanliness and comfort are uniquely American product needs. The cleanliness of the design gave Lucky Strike an American image, which [in turn] gave it a national market. By buying a pack you automatically became an American."

Packaging soon spread from consumer food products to restaurants. The early fast-food hamburger restaurants White Castle and White Tower also used boxlike packages to enable consumers to stack multiple items in a single bag. Eventually, restaurants became packages customers could walk into. The crenellated roofline of a midwestern White Castle and the orange roof of a Howard Johnson were visible from the car, making these establishments stand out from the competition. The concept soon was adapted by all manner of businesses.

One of the most successful grocery store package designs came from Donald Deskey, who created the still-used design for Tide detergent for Procter & Gamble. Using three simple colors—orange, yellow, and blue—Deskey created a new design that referenced past products and consumer associations. The bull's-eye design, created in an undulating pattern, recalled Oxydol, an earlier detergent, as well as the swirling motion of the washing machine agitator. The use of orange and yellow reinforces the theme, because the visual rhythm suggests movement. The outer circle of the bull's-eye extends beyond the boundaries of the box, which historian Hine says "allows shoppers to participate in the design." The colors that dominate—orange and yellow—suggest power, or heavy-duty use. They're the colors of traffic cones and warning signs. The blue introduced in the "Tide" lettering suggests softness and mildness. Blue, in modern package design, means clean. A similar Deskey design, for Cheer, used an opposite color scheme—primarily blue, with streaks of red and yellow. The eye-catching elements were three boldly colored designs that could have been brushstrokes or abstracted sails. Deskey meant them to resemble clothes billowing on a clothesline.

Deskey estimated that the visual treatment of a product accounts for only 13 percent of the time dedicated to designing a product. He allotted 37 percent for market research, 40 percent for technical materials research, and 10 percent for client management. His Procter & Gamble package work for the dishwashing soap Joy also creates a great visual image coupled with subtle design motifs that suggest emotions to the purchaser. The product name is spelled in all capital letters, each in a bright contrasting color, suggesting the product is a joy to use. The contoured bottle has a stylized hourglass shape with a ribbed surface, referencing a feminine form and not coincidentally making the container easier to grip, particularly with soapy hands. Finally, the carton containing the product was somewhat larger than the bottle, making it seem like the customer was getting more for the money. Deskey's 1962 design for Micrin mouthwash is one of the designer's wittiest creations. The bottle is shaped like a flask with a screw-on cap shaped to resemble a stopper. This modern update on classic apothecary bottles of the nineteenth century brought out nostalgic associations that significantly increased the product's sales.

The explosion of product packaging was mirrored in almost every other postwar product category—furniture, home appliances, textiles, and so on. The Knoll and Herman Miller furniture companies transformed the American kitchen and living room by popularizing modern designs using steel, plastics, and plywood. Charles Eames's iconic lounge chair, Eero Saarinen's Womb chair, George Nelson's "marshmallow sofa," and modular storage systems were installed in homes in both "style-conscious" centers such as New York, Chicago, and Los Angeles, and in everyday American cities like Omaha, Bismarck, Memphis, and Toledo. Kitchen appliances were undergoing a sea change as well. Stoves were morphing from stand-alone white cubes to gleaming stainless steel cooktops that could be installed in new Formica counters. Refrigerators were appearing in places other than the kitchen. The "Wonderbar" mini-refrigerator by Sevel was designed to store refreshments in the living room or home bar. General Electric developed a "horizontal" XR-10 refrigerator, designed to be installed above a counter. The prototype was greeted with great enthusiasm, but unfortunately many of the same consumers were unwilling to change the layout of their existing kitchens to accommodate it.

Home products were also changing colors from antiseptic white to black to more adventurous shades such as copper, avocado, turquoise, and the now-dreaded "harvest gold." Designers such as Loewy, Eva Zeisel, and Russel Wright brought modernist looks to ceramic tableware that could be purchased inexpensively and marketed at department stores.

The exponential expansion of areas in which designers chose to specialize meant that most design firms could not offer services addressing every need. The original designer-entrepreneurs sold themselves and their services by attraction and recognition. But the designer as impresario was fading into the background—to be replaced by the Man in the Gray Flannel Suit. This "organization man," especially one who performed design functions, was much more interested in finding a salable product, not necessarily the most advanced or most groundbreaking product. More manufacturing companies began to bring their marketing departments into the design process, which effectively negated Loewy's dictum to limit the number of people with the ability to approve a product. In fact, several designers veered away from pure product design to become market researchers.

The marketplace also was becoming more independent. As the nation emerged from World War II and established itself as a military, political, and financial powerhouse, national confidence grew as American consumers began to believe in the inarguable correctness of their own taste. Though there were still countless millionaires plundering the destitute estates and moneyed families of Europe for perceived "good art," there were a larger number of consumers who wanted to buy well-designed products and dazzling art created by American artists. During the war, manufacturing of consumer products had been severely limited. The factories that had pumped out war materiel in quantities so large that it dwarfed the production of both the country's allies and the enemy powers were ready to convert those facilities back to consumer products.

As THIS SEEMINGLY ENDLESS market appeared, fueled by the postwar population boom, demand far outstripped the capacity of industrial design firms. As everyday appliance companies looked to their employees to design their wares, the independent industrial designers looked beyond packaging and sought new arenas where they could sell their expertise. The advent of popular airline travel in the 1950s opened a new niche for designers to conceive aircraft interiors and exterior decorative motifs. Henry Dreyfuss set up a branch office outside the Lockheed Aircraft factory in Burbank, California, and Walter Dorwin Teague opened a branch office in Seattle, Washington, to service Boeing.

Dreyfuss created the interiors for the Lockheed Constellation, perhaps the most beautiful of the 1950s propeller-driven airliners. The Constellation was divided into compartments, including a lounge with divans and map murals. For Dreyfuss, the Lockheed assignments reached their climax in the Electra, an airliner with an artfully conceived exterior. The interior was impractically beautiful. The compartments were created as living rooms, with tables, lamps, and armchairs. Overhead storage compartments were eliminated to give passengers more headroom. The design seemed to be conceived in an alternate universe where airliners never declared emergencies or crashed. It is hard to imagine passengers leaping out of armchairs and pushing aside table lamps to escape during a runway emergency. The interior captured the optimism of the 1950s, where technology would never fail and travel would occur in luxurious settings.

The fledgling airlines initially charged premium prices to serve the limited numbers of passengers able to afford their services. Designers were asked to create lavish flying living rooms, complete with private seating areas, comfortably plush seating, and luxurious detailing. Teague's company set the standard for today's airline design in its work for the Boeing 707. Instead of designing a mockup of a cabin section, Teague and his designers created a detail-perfect real-size model of the entire cabin. In extensive testing, the designers simulated flight conditions, including full service and sound effects, and they subjected test passengers to actual flight times between different cities.

Teague's company also was the first to make significant inroads in creating designs for clients in the federal government. Loewy, Dreyfuss, Teague, and others had taken on commissions for various federal agencies during the war, but none had established a long-term relationship with a federal or military client. In 1958, Teague signed the largest design consultancy contract in history—worth more than $70 million (around $597.4 million today)—with the US Air Force Academy to design interiors and equipment for its extensive campus in Colorado Springs, Colorado. Teague Associates' good fortune brought its own set of trade-offs. The size of the Teague commission meant that designers were in effect asking to be organization men. With millions of dollars at stake, design decisions were more likely to be safe and conservative rather than innovative. In previous decades, Loewy had answered only to upper management (or even just one man) for Studebaker, the Pennsylvania Railroad, or Frigidaire, as had Dreyfuss with Bell Telephone and Teague with Eastman Kodak. With federal contracts, designers often had to obtain approval through many layers of decision makers. At the same time, client approvals were getting much more complicated thanks to bureaucracy.

THE ROCOCO EXCESSES of the auto industry were mirrored by other consumer product manufacturers, particularly in home appliances. The intensive competition between manufacturers also made it difficult to keep quality levels high. Mimicking the automakers, who worked to keep sales up, appliance companies added features to their products. Refrigerators gained separate freezer compartments, produce crispers, and expanding color palettes. Stoves offered oven lights, griddles, timers, and storage. Advertising for these appliances also borrowed inspiration from car ads. Consumers could aspire to the ultimate kitchen with the same fervor they brought to the desire to own a Cadillac. The large manufacturers such as General Electric and Westinghouse often advertised products in "dream kitchens" designed to make consumers feel their serviceable appliances were somehow inadequate. Eventually, appliance companies began to market specialized products such as electric skillets, popcorn poppers, and blenders to boost demand for new products.

Loewy and Associates experienced this home appliance sea change in its commissions for Frigidaire. The Loewy firm created designs for the GM-owned company from 1939 to 1954. (Loewy claimed he commuted from New York City to Dayton, Ohio, for sales calls to the company for more than twenty-five years.) By the late 1950s, the rounded refrigerator designs pioneered by Loewy and Associates looked jarring in a space-age kitchen. In addition, the introduction of TV dinners and frozen food products made the earlier incarnations of refrigerators, which featured internal, often ice-clogged freezers, obsolete. In 1954, the Loewy firm presented Frigidaire with redesigns for the company's entire line of nine refrigerators for the 1957 model year.

Frigidaire, unhappy with the Loewy concepts, opened the contract for bidding. GM's industrial design firm won the contract by designing the entire line from the ground up. The result, which received and deserved high praise from consumers and design critics, was the "Sheer Look." The appliances were minimalist and unlike anything on the market. The nameplates, handles, and hinges were either hidden or incorporated into the refrigerator's chrome framing. The door was set inside the cabinet so that the door seal didn't show. The new look could fit flush with modern cabinetry, which made Loewy's design appear dated.

Loewy's firm was asked in 1950 to incorporate high design into another dining niche—fine china and porcelain. Rosenthal, a European china and porcelain producer since 1880, had been taken over by the Nazis in 1934 and returned to family control after World War II. The company made the decision to hire artists and designers to execute original designs for their china. Owner Phillip Rosenthal said, "Throughout history, works of art and purely functional objects retain their cultural and material value only when they reflect the spirit of their age." He was as good as his word. After the war, the china company occupied a new factory building designed by Walter Gropius.

Loewy was hired to ensure that china and glassware were not treated as heirlooms gathering dust in the dining room cabinet but rather items to be updated every few years. Loewy introduced four new lines of china in the late '50s, including Continental, which remained in production until 1975, and the Form or Shape 2000 line, which was stocked in stores through 1978.

Most of the Rosenthal china was reimagined by Loewy employee Richard Latham, a product designer in the Chicago office. Latham came

As more US companies began to hire in-house designers, Loewy and his staff turned to international companies for commissions. In the 1950s, Loewy and Associates created stark, geometric china for Rosenthal, a German manufacturer of fine china and dinnerware. (Raymond Loewy™/® by CMG Worldwide, Inc./www.Raymond Loewy.com)

out of what was the first corporate design shop, the Bureau of Design, for Montgomery Ward. Created by Anne Swainson, the Bureau of Design was put into operation to redo the famous Ward's catalog. Swainson switched the look of the catalog from antiquated woodcut illustrations to perfectly composed, inviting photographs. Swainson oversaw thirty employees, most of them trained to be her product designers. Latham was hired by Loewy in 1945 and immediately made a splash with his design for the military-inspired Hallicrafters shortwave radio.

Somewhat surprisingly, Loewy continually singled out Latham as the Rosenthal line's principal designer in publicity materials. Latham and Loewy also created a minimalist molded plastic line of dinnerware for Rosenthal's Lucent Corporation. The dishes, which came in white, pink, turquoise, and yellow, were thin and nearly as translucent as porcelain. Although the firm's "finer" Rosenthal designs received more praise from the public at the time, the Lucent dishware remains a simple, elegant, and beautiful line of inexpensive mass-market dinnerware.

Although Loewy's firm was still profitable as the 1950s ended, partially because he used his European roots to bring in international clients, the potential pool of commissions kept evaporating. At the same time, more product markets were created by the influx of disposable income. Televisions were starting to appear in many homes. Interestingly, few independent designers attempted to create a new look for television sets, and most manufacturers used wooden cabinets, in effect disguising the set as a piece of furniture. Television sets, save for one or two sci-fi designs, resisted innovative design until the turn of the twenty-first century and the debut of flat-screen and plasma televisions.

Other markets that had been previously ignored opened up, such as sport boats and outboard motors. Major designers entered this market, Loewy with Dorsett Marine, and most notably Brooks Stevens, who designed sleek, beautiful outboard motors for Evinrude. Stevens was straightforward about his design aims. As leisure time continued to expand, some of the blue-collar equipment that previously might have been housed in a factory started to appear on home workbenches. Industrial tools such as saws and drill presses were adapted and designed for home use. The largest new market for designers was the niche where Loewy started his career: office equipment. Eliot Noyes was hired by IBM to create not only such benchmark designs as the Selectric typewriter, but also the company's entire corporate identity, which included products, office décor, and corporate stationery. The design world soon started to tackle copiers, projectors, portable presentation screens, and the large cabinet-style mainframe computers made famous by IBM.

Office furniture emerged as the 1950s niche with the most far-reaching influence. The gold standard of office furniture was found at Herman Miller, a Michigan-based company that came to prominence in the 1930s with contemporary designs by Gilbert Rhode. When Rhode died in 1946, Herman Miller brought in George Nelson. Nelson recruited a series of influential designers, including Charles Eames. Nelson, best known for his sling sofa and marshmallow sofa (a modernist design in which the sofa back is composed of circular cushions), kept the company on the cutting edge of design for decades. Eames, who eventually formed his own design firm with his wife, Ray, created a series of seminal designs, including the classic formed plywood cafeteria chair, a storage system that used a metal frame to hold wooden drawers, and the "Eames chair," a formed laminated rosewood lounge chair and ottoman that featured rich leather upholstery. The Eames chair still is a major seller for Herman Miller (and myriad other companies that produce knock-offs of the design). Eames, who started his career in a corporate setting, was perhaps the last of the major designers to establish an independent reputation as a designer for hire.

As DESIGN HISTORIAN Jeffrey Meikle noted, designers moved "From Celebrity to Anonymity" during the 1960s. Norman Bel Geddes, who blazed the trail in the 1920s, was out of business by the 1940s, a victim of ego, lack of business acumen, and a growing indifference to "genius designers." He died in 1958. Walter Dorwin Teague and Henry Dreyfuss maintained practices into the 1960s, but they found their influence dwindling. Loewy was able to stay in the public eye into the '60s, largely through his growing federal commissions. Another factor in Loewy's record as a nearly permanent figurehead for his profession was his long life. By 1970, Loewy was the only designer of the original Big Four still living—producing steady, albeit declining, output.

When the seminal designs of the 1930s came back into the public consciousness in the 1970s, Loewy's hammered-in-steel training in public relations reemerged to serve him well. Loewy was the sole surviving authority available for historians, journalists, and critics. Loewy's unbroken string of designs over fifty years made him during those years a godlike figure, a persona he was only too happy to cultivate. If in the beginning of the 1930s Bel Geddes, Teague, Dreyfuss, or Loewy could singlehandedly remake entire industries, all were destined to be less influential with the passing of each decade.

As the 1950s reached its midpoint, many of the major independent designers, seeing the scarcity of domestic commissions, looked overseas. Loewy reopened his London office in 1948. But Douglas Scott, who ran the prewar incarnation of the office, by then had established his own design firm where he accumulated commissions from previous Loewy clients and from his own work, most notably the 1954 London Routemaster bus, the famed red double-decker people-mover. Scott also taught a new industrial design course at London's Central School, where one of his students was Patrick Farrell, who would go on to become chairman and co-owner of Raymond Loewy International—the final incarnation of Loewy's company. Unfortunately, the London office was again forced to close, in 1951, because the office could not overcome financial restrictions imposed by the British government's Foreign Exchange Control.

Eventually, the European offices would bring in many more commissions for Loewy, ensuring the survival of his agency into the 1970s and reintroducing his name as a European entity. Of all the renowned American designers, Loewy profited the most from the international offices, but that was due more to the longevity of his career than anything else.

Loewy's foreign office in Paris would become his most important business center. It opened in 1952 to ensure a European presence after the London office failed. The tradition of design in Europe was much more theoretical and intellectual than the design ethic Loewy had helped start in the United States. The "vulgar taste" that had served him so well in America was considered to be just vulgar in Europe. The products Loewy shepherded through the American markets were gauged to sell briskly. The European designers took a more intellectual tack that emphasized philosophy rather than marketing

Ironically, Loewy was seen in France as being too American, according to Evert Endt, who worked as art director for CEI / Raymond Loewy in Paris. Still, the Paris office was aware that America viewed the designer as the ultimate European. While Loewy imbued the office with an American marketing mission, the working atmosphere was organized in the more communal European style. Still, some of the idiosyncrasies that fostered resentments in Loewy and Associates' US offices seeped into the foreign offices as well. Loewy put his name on all work emerging from the Paris office, and most of the staff acquiesced without complaint, viewing the practice as a way of trademarking the firm's work, much like the products generated under Walt Disney's name.

Additionally, the 1950s saw the reemergence of the European design aesthetic—a sleek, minimalist, monochromatic vision seen in the shavers, appliances, and kitchenware designed by Braun and other manufacturers. Within a few years, the European look, no doubt helped by the emerging globalization of the American economy, became the dominant style for the discriminating (or nondiscriminating, for that matter) consumer. Critics and tastemakers, disenchanted with the assembly-line application of design tropes, saw America's designers as chrome-happy decorators incapable of producing a clean design. Part of this disenchantment derived from the industry's failure to found industrial design programs that left a public footprint. New York's Pratt Institute and the Carnegie Institute of Technology were early academies for the profession, but critical influence shifted to two institutions: Harvard University, characterized by an architecture program run by Walter Gropius, and the Illinois Institute of Technology, run by Mies van der Rohe. Both programs were devoted to the "pure" design aesthetic of the Bauhaus. The less-is-more aesthetic came to dominate industrial design starting in the 1960s, but until the International Style of architecture filtered down into everyday design vocabulary, the popularization of the look further marginalized the independent designers.

Toward the end of the 1950s, the movement for "space-age" design also found a willing market of consumers. The International Style pleased the design and art critics, but furniture by George Nelson or Eero Saarinen or a house plan by Gropius was typically only within the reach of the upper class. The space-age design era was initially applied to more affordable goods such as radios, wallpaper, clocks, and toys. The popularizers of this look were Loewy's great rivals Harley Earl and Virgil Exner. General Motors, Ford, and Chrysler allowed car buyers to launch themselves into the jet set by offering them low-slung machines that sported rocket tail fins, chrome-laden bumpers, conical mirrors, and sparkling dashboard panels meant to ape the controls of a fighter or space machine. While the futuristic motifs did not become popular in furniture design until well into the 1960s, corporate designers applied the car stylists' lessons to appliances, producing refrigerators that incorporated futuristic chrome trim as well as washers and dryers that came equipped with controls that mimicked dashboards.

While design consultancies fell from favor, or at least they had to work much harder to find clients, the major firms were still financially able to create groundbreaking designs. Loewy, even when it became obvious that the era of the independent designer was waning, kept touting the designer as a singular visionary. Every essay he wrote and many of his media interviews, including one of his last television appearances on a segment of the CBS television show *60 Minutes*, insisted that designers should not be tethered to a single company or industry. Design historian Paul Jodard described Loewy's "restless versatility" as the basis for his belief, but Loewy felt in his bones that designers should be able to do it all. "Like war, which cannot be left to the care of soldiers, design was too important to be left to the care of company men," Jodard wrote. Ironically, the main beneficiaries of his insight into the importance of design were to be the generation of Charles Eames (initially turned down for membership in the American Society of Industrial Designers because his work as a furniture designer was too narrow in scope) and others who worked closely with individual companies, rather than ranging across the field of design. They owed much of their freedom to the ground rules created by Loewy and his contemporaries.

13

The Long Road Down

By the start of the 1960s, Loewy and Associates was showing significant signs of stress. The loss of the Studebaker account was a body blow for the company. Other clients took up some of the slack, but often these commissions were for specific products in a set campaign overseen by others—strictly turnkey assignments. The flagship accounts that generated more than $1 million a year for Loewy, such as Studebaker and the Pennsylvania Railroad, were by the end of the 1950s on the wane or gone altogether.

Loewy began the '50s by publishing his memoir *Never Leave Well Enough Alone*. Of all the books produced by the industrial design giants—Bel Geddes's *Horizons*, Dreyfuss's *Designing for People*, and Walter Teague's *Design This Day*—Loewy's was by far the most successful. It was also one of the last to be published and served as the first and most convincing step toward burnishing Loewy's legend. Every anecdote, hypothetical situation, or quote has been polished with the rhythm of the well-told tale. Characterized as "a 100,000-word after-dinner speech," by Peter Blake, then associate editor of *Architectural Forum*, the book manages to be funny, entertaining, informative, and egotistical—sometimes within a single sentence. The memoir takes the same rags-to-riches plotline as the most inspirational American fiction, and the easy victories and smooth storytelling that characterize Loewy's design triumphs suggest that his recollections were, if not fictionalized, then entertainingly embellished. The chapter headings, called intermezzos by the author, are presented in differing type fonts, each chapter introduction explaining the design principles illustrated by the typography. He explains the dynamic use of italic type, the juxtaposition of curves and lines, and a host of other informative and relevant design tropes. The writing in each chapter can be highly descriptive and dynamic. There are sections of dialogue and imagery that could be produced as a screenplay—particularly scenes where Loewy meets with Martin Clement, head of the Pennsylvania Railroad, and the scene from Dayton, Ohio, as the CEO of Frigidaire takes the designer on a night drive to the company's plant. Loewy's life story on the page was better than his rivals' because, in order to function as a sales tool for Loewy's work, it was required to be. To paraphrase the old joke, as an American, Loewy had to be more Catholic than the pope. And, like most converts, he was zealous in his pursuit of the American ideal. From his sports cars, dune buggies, western outfits, and laser-focus on work to his intimate understanding of the media of his time, Loewy had left France in his rearview mirror. Design historian Stephen Bayley explains it best: "Loewy was an import. Because he was born in France he would always maintain a particular appreciation of what it was to be American, and this too would be the grit in the oyster that made him something special."

Blake's snarky "after-dinner speech" comment came from his deeply held credo of the modern architect Louis Sullivan: "form follows function." He praises Loewy's success but denigrates his talent, saying he "lives in the special Aura of a grade B movie, and the Box Office is doing okay."

Loewy's memoir is self-consciously perfect—the biography as slogan, according to one design critic. In describing a portrait of Loewy in the memoir (the famous shot of Loewy posing in front of the Museum of Modern Art "designer studio" he created with Lee Simonson), design writer Philippe Tretiak said Loewy was like a "glamorized movie star—too glamorized. Something about the tilt of his head, his mesmerizing fortuneteller's gaze, his initialized cufflinks, casts him in a supporting role. Loewy always had the allure of the Argentine tango dancer, a flamboyant upstart who gave the impression of flashy, frozen elegance." The tango dancer was a born embellisher and raconteur, and he never stopped burnishing his stories or his biography. The media of the day, where journalists' research by necessity rarely reached beyond clip files of major papers and magazines, aided in creating the legend by repeating Loewy's best stories over and over. The Coke bottle claim, the modernist house, the car designs have been polished into fact through sheer repetition. Indeed, Loewy never left well enough alone when it came to selling the client or the reader on his genius. "He was the Leonardo Da Vinci of the New World—both engineer and visionary."

Loewy had begun to edge toward retirement or at least part-time leadership by the time the 1960s drew to an end. He was never a year-round manager, spending long periods traveling even in his early career, but by the '60s he rarely made it to branch offices. He still made important presentations for the firm's major clients and continued to aggressively market himself as the public face of industrial design, but much of the continued success of Loewy and Associates after 1960 can be traced to the retail design operation of Loewy's partner Bill Snaith.

Industrial designers "proved that good appearance was a highly salable commodity," Loewy wrote in *Industrial Design*. They "also opened the floodgates to all sorts of unethical merchants and phony designers who believed that cosmetic camouflage could conceal shoddy products and tickle sluggish cash registers."

As Loewy's sentiments reveal, he couldn't bring himself to refocus the firm's structure, perhaps specializing in retail design and packaging, for single companies. Design historian Thomas Hine called the decade of 1954 to 1964 "one of history's great shopping sprees . . . many Americans went on a baroque bender and adorned their mass-produced houses, furniture and machines with accoutrements of the space age." The shopping spree was funded by a great widening of the middle class. The prewar middle class was composed of small business owners and merchants, usually in large cities. After the war, the employees working in massive factories were earning salaries that plunked them squarely in the middle class. The shop owners of the prewar years expanded their operations in size and location, often establishing branches in newly sprouted suburban communities. Many of these new consumers with additional disposable income were employees of large (and steadily getting larger) corporations.

At the same time, Americans were stepping away from their past. Entire city neighborhoods were whisked away by bulldozers in the name of urban renewal. Modern homes were being reproduced in suburbs across the country. *House Beautiful, Better Homes and Gardens*, and other shelter magazines eliminated older homes from pictorials. Historian Hine referred to this collective rejection of the past as "Populuxe."

Products, possessions, and passions all were centered on the future. There was no reason to hold on to the past. Even modern design was not safe. Harley Earl, looking to refit his office in the GM Design Center, ordered an Eames chair and ottoman and replaced the rich leather with orange Naugahyde.

Henry Dreyfuss recognized that the marketplace had turned away from the "great man" model. He changed the name of his firm to Henry Dreyfuss and Associates in 1967, and by 1969 he had left the firm entirely. He became a sort of senior adviser to his larger corporate clients, and he and his wife, Doris Marks, wrote *Symbol Sourcebook: An Authoritative Guide to International Graphic Symbols*. Shortly after the publication of the book in 1972, Dreyfuss sat with his wife, who had been diagnosed with cancer, in their car and turned on the ignition in an enclosed garage. The couple had worked together for forty years, and collaborated a final time in suicide on October 5, 1972. Loewy had respected Dreyfuss more than any of his other competitors and would outlive his friendly rival by fourteen years.

Loewy had not aggressively pursued government work for much of the tenure of Loewy and Associates, preferring the control (or illusion of control) that corporate commissions provided. He did take on World War II military commissions, when most of America's manufacturing base was dedicated to the war effort. Designers were asked to design products for the military and for making wartime rationing and deprivation more palatable. Beyond the much-referenced lipstick commission, designers in the Loewy offices also produced a habitability study for the US Army (the firm would be hired to complete a similar project for the navy after the war). Another ambitious military design was a glider for the Medical Air Corps that could come in for a landing and immediately be converted into a field hospital. Other designers found lucrative wartime commissions. Even Norman Bel Geddes, perhaps because he was temperamentally suited to working on larger-scale projects, took advantage of military projects. Bel Geddes, maestro of the scale model *Futurama* for the 1939 World's Fair, developed a series of model boats that were used for tracking sea battles on maps and training in ship recognition.

As his corporate accounts dwindled throughout the 1950s, Loewy looked to federal agencies for work and found it not in the halls of Washington, DC, but on a runway in Palm Springs. John F. Kennedy's *Air Force*

One arrived in March 1962. The trip is ingrained in pop culture memory because Kennedy famously snubbed Frank Sinatra—a Palm Springs resident who had redesigned his home to host the president—after his alleged mob connections made the White House staff leery of associating with the entertainer. Loewy, who was nearly as well known as the show business denizens of Palm Springs, did meet the president. The social connections Loewy cultivated allowed him to continue soliciting commissions throughout the 1960s and 1970s. His friendship with the Kennedys (although his connection was more with the style-conscious First Lady than the president), cemented over the design of *Air Force One*, paid dividends well beyond the end of Camelot. The New Frontier commissions gave Loewy instant credibility and entrée into work with federal agencies.

Air Force One has an iconic status in the United States, its colors and shape instantly recognizable all over the world. The president's jet has been linked with memorable images, from Lyndon B. Johnson taking the oath of office on the plane after the assassination of John F. Kennedy, to any number of shots of presidents descending the stairs after a triumphal trip overseas or campaign stop. The aircraft is a place where policy is made, friendships are forged, and seminal events in history occur. Such a vehicle carries weight as a symbol, which means, and meant even more so in Loewy's time, that *Air Force One* carried the force of the United States in its appearance.

Harry Truman, who loved to fly and flew so often that some airline historians credit him with popularizing air travel, named his own aircraft, a four-engine Douglas C-118, the *Independence*. The plane survived an air force effort to name the plane *The Flying White House*. Truman also introduced a presidential paint scheme, featuring a blue fuselage, and personalized the appearance of *Independence* by painting the visage of an eagle on the nose of the aircraft.

Dwight Eisenhower inaugurated presidential jet travel, but his first plane was a prop-driven Lockheed Constellation called *Columbine II* (named after his wife Mamie's favorite flower, *Columbine* also was the name of the aircraft he used in World War II). The plane's color scheme, with little thought to patriotic imagery, was gray, blue, and green. Eisenhower's plane, however, was the first to be called *Air Force One*.

John F. Kennedy was the first president to have an aircraft dedicated exclusively for his use. He loved the name *Air Force One* and decreed that the name should be used in public and in White House publicity. Kennedy realized that the president's plane in some ways represented the power and might of a nation that could deliver its leader anywhere in the world. His *Air Force One* was a Boeing 707, a passenger jet that remained in use as the presidential aircraft until 1990, when President George H. W. Bush upgraded to a 747.

Jacqueline Kennedy, whose design sense cut a wide swath, hired Loewy to give *Air Force One* a makeover. In *Never Leave Well Enough Alone*, the designer described a long design session with the president sitting on the floor of the Oval Office. In his book *Industrial Design*, Loewy makes the claim that the commission came directly from John F. Kennedy, which is almost certainly not the case. He also claimed that his last great commission, the interiors for the National Aeronautics and Space Administration's Skylab, came as a result of the presidential airliner commission, but the NASA commission came nearly five years after *Air Force One*, and there is no clear connection between the two projects.

According to Loewy, Brigadier General Godfrey McHugh, the air force aide to the president, suggested that Loewy contribute design ideas. Loewy was well acquainted with McHugh through his friendship with Air Force Secretary Stuart Symington. A new *Air Force One* was being developed, and McHugh suggested that Loewy redesign the markings. Loewy claims he did the proposal design for no pay. "I received a phone call from [McHugh] telling me that the president wished to meet with me. I flew to the White House, the beginning of a remarkable relationship." Loewy came to the meeting and returned a week later with some sketches, showing Kennedy four different looks. "In every case I had replaced red [the predominant color of the jet] with a luminous ultramarine blue."

According to 1967 notes in Loewy's archive in the Hagley Museum and Library, the designer came to the White House with four graphic proposals and five lettering ideas. According to the notes, Loewy's first ideas were predominantly red. Kennedy, according to the notes, made the decision to make the color pattern blue. According to design historian Phil Patton, Secretary of Defense Robert McNamara had to personally authorize the color change.

Loewy goes on in *Industrial Design* to describe being sprawled on the floor of the Oval Office with the president, cutting paper and pasting shapes onto sketches. The scene is hard to believe, mainly because

Kennedy's severe back problems are well documented and the scene is too reminiscent of Loewy's description of his meeting with tobacco executive George Washington Hill. In the Hagley notes, Loewy is the one on the floor; the president sat in his rocking chair observing.

The previous version of *Air Force One* featured an abstracted red sweeping cowl offset by a nose painted flat black. "I was unimpressed by the gaudy red exterior markings. The plane's main section was white on top and silver below [the color line started just above the windows]." Loewy's final design retained the white top section but substituted sweeping shades of blue, including the aquamarine that covered the lower nose section and the cowling on the engines. After Loewy and the president chose a blue color scheme, Kennedy asked his secretary, Evelyn Lincoln, and the First Lady's social secretary, Mary Gallagher, to pick their preferences. They opted for the blue prototype as well.

According to Patton, White House appointment diaries do not have records of a Loewy visit on May 8 or May 15, 1962, although there is a record of a long lunch break. According to White House historians (and considering Kennedy's active womanizing), not every appointment was recorded. Although Jackie Kennedy was socially acquainted with Loewy, she is not mentioned in his meeting notes, but several other accounts have her contributing design ideas to the project.

Previously, presidential planes had been identified by "United States Air Force" or "Military Air Transport Service" along the upper fuselage. Loewy substituted "United States of America" on the fuselage and placed a flag on the tail section. The flag's union (the blue section) faced toward the nose. Loewy dialed back the intensity of the blue in the flag in order to more closely match the blue used on the rest of the jet. The typeface for the fuselage lettering was supposedly inspired by the type used in the heading of the Declaration of Independence during a visit to the National Archives by Loewy or a staff designer. "United States of America" was set in widely spaced Caslon typeface.

Loewy added the presidential seal at Kennedy's request, although General McHugh pointed out that the seal should only be shown when the president was on board. Kennedy made a few calls, among them to the secretary of the air force, to get approval for the change. Loewy also was asked to design the presidential stateroom. Loewy recounted that the president asked him to design a rug that would lie between twin beds. The rug's design was pale blue with an eagle in the center of an oval

Loewy's redesign of *Air Force One* gave the presidential airplane its distinctive blue-on-blue color scheme complemented by the singular speed line bisecting the color sections. Loewy also suggested placing "United States of America" on the fuselage. (Raymond Loewy™/® by CMG Worldwide, Inc. / www.RaymondLoewy.com)

formed by thirteen stars. Loewy asked a friend, Edward Fields, owner of a carpet-manufacturing firm, to weave two rugs over the weekend. Eventually, all VIP aircraft used at Andrews Air Force Base used the same Loewy color pattern as the presidential aircraft.

Loewy's Kennedy connection continued after the president's death. Jacqueline Kennedy asked Loewy to design a commemorative stamp, which was to be issued on May 29, 1964, the president's birthday. The stamp featured a double panel showing a blue-screened version of a snapshot selected by the former First Lady next to a representation of the eternal flame at Arlington National Cemetery. The text reads, "And the glow from that flame can truly light the world."

Loewy's aircraft-related designs could be considered his least celebrated works, but in their time, his designs for *Air Force One*, Trans World Airlines, Air France, and others were visual shorthand for global businesses recognized by millions. Loewy's work for TWA is referenced in the 2005 film *The Aviator*, albeit briefly. In the film, Howard Hughes, deep in the throes of paranoia, tells one of his factotums to "fire Ray Loewy" because he suspects the designer of leaking designs to Pan Am executive Juan Trippe. Although it's doubtful Loewy ever answered to being called "Ray," his influence on Hughes's airline was significant. In

1959, when TWA rolled out its first workhorse jet, the Boeing 707, the design firm was asked to create a logo. The designers delivered a logo that was the most recognizable aviation symbol of its time. It featured a red TWA slanted backward, as though flying through a headwind. The letters were superimposed over golden interlocked globes. Loewy's designers placed the logo on the tail and above the boarding doors. But what made the design work was a red stripe, stretched the length of the fuselage, that slightly widened as it traveled from cockpit to tail, giving an indelible image of velocity. "TWA pilots were said to boast that their jets seemed to be going 600 miles per hour even when parked," wrote a *New York Times* reporter. The twin globe logo was eliminated in the 1970s, when the airline's global image attracted international attention from hijackers and terrorists. Soon after, the airline shed its international routes when corporate raider Carl Icahn took control of the airline. Loewy's simple visual shorthand for international air dominance was subsequently lost in the clouds, absorbed and eliminated from existence.

While federal commissions sustained Loewy's American operation for a time, most of the lucrative incoming work after 1960 was coming to the company's European branch. One of the largest clients for CEI, the firm's Paris operation, was Shell Oil. Loewy did not originate the company's abstracted shell logo, but his firm redesigned the emblem, removing the word "SHELL" from the interior of the pectin shell and instead placing it below the emblem. CEI also developed uniforms for station attendants. The work clothes seem heavily influenced by the Carnaby Street "mod" look.

Though Loewy was hardly the first designer employed by oil companies, he was perhaps the most prolific. Loewy's work for Standard Oil of New Jersey, then known as Esso, resulted in one of his most memorable corporate identity commissions. The company contacted Loewy in 1966 to let him know that it had decided to make a worldwide change of name. Loewy was sworn to secrecy for the project, code-named "Nugget" within the designer's office. Loewy took the commission and, according to *Industrial Design*, he went to his Palm Springs studio at *Tierra Caliente* and came up with the interlocking double-X logo. He says as much in a letter, telling Exxon executive Francis X. Clair his memories of designing the logo. "I prepared in my personal design studio about eighty rough pencil sketches." In the designer's archives at the Library of Congress, there is a slip of drafting paper with sample sketches of the Exxon logo delivered

As Loewy and Associates' automotive and product design operations waned in the late 1950s and '60s, the company moved into creating corporate identities. Loewy's most extensive work in this area came from the Shell Oil Company, but Loewy's interlocking Exxon logo was more widely praised at the time and in the present day. (Raymond Loewy™/® by CMG Worldwide, Inc. / www.RaymondLoewy.com)

to the client in 1966. A handful of brisk sketches fill the page. There is one circled with interlocked X's, which suggest a crossroads. Loewy preferred the design because it mirrored the two S's in Esso.

Despite the wane of manufactured product commissions, corporate identity assignments like those from Shell or Exxon emerged as a business niche. Loewy was able to use his acclaimed work for International Harvester and his work on *Air Force One* as a calling card for other identity commissions. Companies that by the 1960s created their own product design in-house often lacked the staff or creativity to market the firm as a company. Loewy was just one of the designers to exploit this rapidly expanding market. One of Loewy's least-known (but most often seen) federal commissions was the 1970 logo for the US Postal Service. Commissioned when the federal agency became an independent corporation, the deceptively simple design, which combined a silhouetted eagle over the "US Mail" logo lasted twenty-three years until Postmaster General Marvin Runyon ordered a redesign in 1993. Today's design, which Runyon admitted "evolved" from Loewy's work, was created by CYB Yasumura Design, a subsidiary of the advertising agency Young and Rubicam. The Yasumura redesign was selected from three hundred different designs and went through a dozen focus groups before becoming the final logo. The transition from the "great man" design studios to the faceless corporate designers dependent on market research is a stark contrast that perfectly defines each era.

In *Industrial Design*, Loewy claims the *Air Force One* commission led to a much more ambitious program in which Loewy would apply "the

Loewy began to work for federal agencies after creating the look for *Air Force One*. He went on to design the logo for the US Postal Service in 1970, perhaps the designer's best work for corporate identity. The design lasted twenty-three years. (Raymond Loewy Archive, Courtesy Hagley Museum and Library)

applications and concepts of industrial design to some of the important issues of the period, including highway congestion, urban blight and redesign of government buildings." This assignment may have come up from Loewy's own suggestion. He and his firm were known for aggressively suggesting new projects once they had been asked to participate in the initial commission. The White House formed a team to look at the proposal; Arthur Schlesinger Jr. was identified as Loewy's administration contact. If there were any surviving proposals from any of Loewy's conceptual solutions to these massive urban issues, none have seen the light of day in any of the designer's books or museum exhibitions.

Loewy's ultimate federal commission and his last professional triumph came at the end of the 1960s through his designs for Skylab, NASA's first attempt at a manned orbital space station. Launched on two gigantic Titan rockets on successive days, the cylindrical lab went into space, followed the next day by a three-man crew. At 106 feet long and weighing 100 tons, the "baby space station" was designed, once in orbit, to unfold its wings, resembling "a flying Dutch windmill." The wings were solar panels, generating the power needed to run the space vehicle's instruments and systems. The lab carried thirty-seven cameras, a metabolic laboratory, and data collection devices such as magnetic tape, digital recordings, and film strips.

Unfortunately, Loewy had nothing to do with the designs described above. He was called in after the fact, when most of the scientific design commissions had already been made. In *Industrial Design*, Loewy took great pains to link the futuristic display he created at the 1939 World's Fair to his later work with NASA. The most memorable of Loewy's fair displays was the animated model of space travel, popularly known as the Rocketport. The port consisted of a gargantuan cannon supported by concrete bunkers. A magnetic crane loaded the spacecraft into the cannon. Every twenty minutes, visitors boarded in total darkness except for the spotlighted rocket on its launch pad installed on a shallow pit on one side of the space. The launch area featured blinking white and colored lights. The whole area vibrated and pulsed through the whirring of powerful motors, compressors, and high-frequency sound waves.

A relatively rare shot of Loewy and a workman assembling the Rocketport display at the 1939 World's Fair. Loewy later claimed that such designs for space travel fantasy helped secure his bid to work on NASA's Skylab, but this tableau is more influenced by Buck Rogers than by serious aeronautics. (Raymond Loewy Archive, Courtesy Hagley Museum and Library)

"The rocket seemed ready for lift off and the audience was on edge," Loewy wrote. "Then at the sound of sirens: Lift off! In a moment, a blinding flash of hundreds of strobe lights and the roar of compressed air suddenly released—the rocket, through optical illusion, seemed to disappear overhead in the blackness of space. People who saw it once often returned; it was a thrilling preview of what many believed could happen in the future." Clearly, Loewy's conception of space travel was more influenced by Buck Rogers than any vision of trips to distant moons or galaxies.

At NASA, space missions were beginning to suffer from media over-exposure at the end of the Apollo program, making some lawmakers leery of investing in a floating space station, no matter what the po-tential research value. The initial NASA idea for Skylab emerged from the Apollo program. NASA's chief rocket scientist, Werner von Braun, outlined a series of space project proposals, including space tugs, moon colonies, and shuttle flights between planets. Skylab was chosen as the most feasible.

Budget hawks were more vigilant on the Skylab program, so thrifty NASA engineers reused components from other programs, including launch rockets and docking spacecraft from Apollo, and launch pads, space suits, and hatches from the Gemini program. Even with this frugal-ity, the project's budget totaled $2.6 billion. The lab itself was fifty times larger than the first manned moon capsule, comparable in size to an efficiency apartment. The lab proposal included a bathroom larger than those found on luxury airliners, a kitchen with a large table for family-style meals, and a "porthole" for literally watching the world going by.

The vessel carried three different types of "space" balls that astro-nauts used for exercise. It also carried a "space dart board," which had Velcro to make the darts stick. One of the lab's most famous features was the space toilet. Previous space flights had no bathroom facilities, prompting Apollo astronaut James Lovell's response to a reporter's ques-tion on how he liked the trip: "How would you like to live two weeks in the men's room?"

Loewy's goal, simply put, was to take the bathroom out of the ex-perience of living in space. From the beginning of the space program through the end of the Apollo program, industrial design was perhaps next to last (surely it outranked color choices) on the space agency's list of priorities. With brief space flights, it didn't matter what the capsule looked like or whether the instruments were easy to use. All that mattered was that all the systems worked without fail. In addition, NASA's motto could have been stolen from Loewy: "Weight is the enemy."

The agency proposed two options for a space station. The first option asked NASA to reuse the rocket's final-stage fuel tank as a reused lab. The second option installed a ready-made laboratory in an empty fuel cell and launched it into orbit. NASA chose the latter idea.

The agency came late to the idea that Skylab should be designed with comfort and practicality in mind. Engineers in NASA's Houston, Texas, operations center supported the idea of providing amenities in the capsule, rather than making the astronauts live in an environment tantamount to the inside of a culvert. The engineers in Huntsville, Alabama, where the station was being built, were less enthusiastic about providing amenities. The astronauts of the previous space missions were no fans of NASA interior design, however. Michael Collins, pilot of *Gemini 10*, compared the spacecraft to the front seats of a Volkswagen Beetle, perhaps one of the most Spartan automobile interiors ever designed.

In the fall of 1967, George Mueller, head of the Office of Manned Spaceflight, was so appalled at the impersonal look of Skylab's interior that he suggested hiring an industrial designer to make the space more livable. "Nobody could have lived in that thing for more than two months. They'd have gone stir-crazy," Mueller said upon seeing the first mockup of Skylab. Martin Marietta, the building contractor for Skylab, commissioned proposals and hired Raymond Loewy / William Snaith Inc.

Loewy was 75 years old when the NASA commission came in. Loewy was energized by the opportunity, throwing himself into the commission. He and agency vice president Fred Toerge toured all the subcontractor sites, ending with the facilities in Huntsville, where they received briefings on the entire program and looked over what had been done to that point. The firm produced an extensive report that caught the attention of several NASA administrators, particularly Caldwell Johnson, chief of spacecraft design in the Advanced Spacecraft Technology Division. Johnson felt that habitability had been given little consideration. He brought the plan created by the Loewy designers to Huntsville, where it was met with skepticism and hostility.

After Loewy and company delivered their designs, the engineers were slow to act on them. The Huntsville engineers, who had never worked closely with astronauts (still perceived as knights of the sky, much as in the book *The Right Stuff*), deferred to the flight crews on issues such as color schemes and design. The astronauts, on the other hand, were more concerned with efficiency than appearance. They preferred a workspace where they could do their jobs with a minimum of distractions. The astronauts were less interested in habitability than utility.

Raymond Loewy touted the project in all his speeches and marketing materials. The commission dominates Loewy's book *Industrial Design*; its cover photograph shows Loewy posing in the middle of his Skylab mockup. Although the number of completed NASA designs did not in any way equal the number of proposed designs submitted to NASA, Loewy saw the NASA work as the culmination of a great career. "My first opportunity to express a deep interest in the future of spatial exploration occurred thirty-seven years ago, at a time when such matters were generally ignored. It happened at the New York World's Fair. I designed and staged in the Chrysler Building the simulated launching, in a scaled-down spaceport, of a huge rocket intended for 'international transportation of passengers and mail.' . . . It left the spectators deeply impressed; it was dramatic and a hit of the World's Fair."

A mockup of Skylab's interior, Loewy's last major design assignment. He created dozens of designs on "spec" for the aeronautics agency, but just a handful of his work made it to the final version of the original NASA space station. (Raymond Loewy Archive, Courtesy Hagley Museum and Library)

It may have been the "hit of the World's Fair" (actually, Norman Bel Geddes's *Futurama* was the hit of the fair), but Loewy was hardly the first person to design a rocket ship. In fact, in the late 1960s, journalists and consumers were starting to question Loewy's off-hand embellishments and impromptu claims, and Loewy began to pay the price for his exaggerations through media indifference. The Loewy group was assigned to deliver "comments and recommendations" for the Wet Lab version of the lab. The space agency's account of Loewy's work is markedly less flowery than Loewy's remembrances in *Industrial Design*.

In *Industrial Design*, Loewy quotes NASA's Mueller in a letter:

> You and your organization played a crucial role in the latest of these momentous steps that man is taking to the stars. I do not believe that it would have been possible for the Skylab crews to live in relative comfort, excellent spirits and outstanding efficiency had it not been for your creative design, based on a deep understanding of human needs, of the interior compartment of Skylab and the human engineering of the equipment and furnishings which the astronauts used. That design and engineering applied, in turn, to our follow-on space stations has provided the foundation for man's next great step—an expedition to the planets.

The 1974 letter says a lot without saying much of anything. Loewy tells a vibrant tale of the commission in his final book. He describes taking part in astronaut tests, donning spacesuits, and learning astronaut lingo, which Loewy dubbed "orbit talk," but most of the NASA documentation is written in the bureaucratic memo style of the federal government, not the grateful thanks of a dazzled client.

Once hired, Loewy bombarded NASA and Martin Marietta with suggestions for Skylab designs. Most of the memos and designs were dutifully responded to and filed away, but the space agency accepted only a few of the firm's suggestions. In *Industrial Design*, Loewy gives the reader the impression that all the designs were wholeheartedly accepted. The designs included in the book encompass not only Skylab, but also a space station and base station that the space agency had not commissioned. Loewy posed for the cover of *Industrial Design* magazine after receiving the commission. The photograph shows a snowy-haired Loewy posed in front of a Skylab mockup where viewers can see the galley, galley table, storage cabinets, and a space "porthole"—all spaces Loewy's firm helped to design.

Loewy and Associates designers visited several areas where Skylab contractors worked and submitted a four-page report in 1968. Loewy's introduction shows his business edge was slipping. It said,

> For 35 years I have been privileged to work in close collaboration with top management of approximately 70 percent of the nation's largest corporations. In this situation I have met the technological leaders of America's industry. Men admired throughout the world. It is in this context that I and my team set out from Denver last December fourth to visit the various centers of space activities. That particular Monday turned out to be the most meaningful of my entire career. During the week I met men whom I consider to be the most transcendental intellectual and technological geniuses of our time—possibly, of any other time.

After such a non-content-rich buildup, Loewy criticized Skylab's inadequate lighting, lack of color variety, noisy fans, and grid-like partitions that Loewy's report characterized as looking like a jail. Mention of a window, which later became a centerpiece of Loewy's claims for the NASA work, is mentioned only obliquely. "We can imagine that crew members, feeling encaged in the OWS [Orbital Workshop] might become excessively obsessed to return to the comparatively cozy atmosphere of the command module, the only room with a view!" The Gemini and Apollo spacecraft had portholes, so Loewy was by no means a window pioneer, but when he was asked to present design plans for the Dry Lab version, his window idea had become policy. The bulk of the report recommended that better living conditions could be achieved by efficient use of space and improved lighting and color schemes. Most of his comments were met with indifference by the agency.

Astronaut Jack Lousma, who went on to command the third space shuttle mission, wrote a letter to thank the designer, a prize treasured by Loewy: "The porthole which was added at our mutual insistence was a great asset to our operations and will be increasingly so for the last missions as they have an extensive program of earth observations to carry out. In fact, our recommendations [sic] for future spacecraft is that they have more windows and larger areas, preferably with a bubble-type configuration."

Once he received the contract, Loewy littered the agency with more queries for a multistory space station and a space shuttle of sorts. Loewy won a nine-month contract to continue work on both options for Skylab, and when the Dry Workshop option was chosen, Loewy received more work from NASA.

Loewy and his designers influenced the interior layout of the vehicle, and many of their ideas to improve crew privacy were adopted. The firm also created a two-piece work uniform. The astronauts fought to retain their one-piece pilot's jumpsuits, but Caldwell Johnson won the argument that the complicated uniforms would be a liability in such cramped quarters. The astronauts did win one battle concerning their jumpsuits. Loewy designer Fred Toerge's uniforms were missing a pocket on the lower leg. Lower pockets are crucial for pilots trapped into a cramped jet cockpit because they are strapped into a confining ejector seat and cannot easily dig into upper pockets; less so for astronauts floating freely in zero gravity for months. Loewy's soothing color schemes also were incorporated. The installation of the window was considered a huge victory for Loewy, and it was one of the few improvements that astronauts reacted to with nearly universal praise.

According to historians, of the nine projects mentioned by Johnson, just three appeared to be accepted: space coveralls, a storage module for uniforms, and the spacecraft's color scheme. Loewy omits the details of Johnson's acceptance letter, so it appears that all the NASA designs Loewy describes in *Industrial Design* were okayed by the agency. A close reading of Loewy's published account of the Skylab design adventure offers a peek into his easily adjusted ethics. Loewy does not mention that the Skylab designs were completed by Raymond Loewy / William Snaith Inc. In the book, Loewy gives the impression that he took the lead on all the designs. After Bill Snaith died in 1974, Loewy felt no obligation to credit him in the 1979 book.

Industrial Design features a variety of sketches that the company submitted to NASA as suggestions for future products and even includes a painting that Loewy executed of a solitary astronaut on the moon. The image, which has little of the flowing life of his fashion illustrations, is not identified as part of a commission.

The work completed for NASA went through extensive testing and debate, but in the end NASA's report gives the company an endorsement. "It is fair to say that we have received sound, professional support from the Loewy-Snaith Company." It was the last of Loewy's publicity coups. In a design career that had celebrated speed and transportation, a commission for space travel was a professional victory that allowed Loewy and his ego to slip the surly bonds of earth. The NASA work also allowed Loewy to grab a national spotlight for the final time—at least in relation to a professional commission.

LOEWY'S FINAL HURRAH in sweeping design influence came in an arena that few of his contemporaries anticipated: product design in the Soviet Union. For a man who bristled at executives and bean counters adding to and subtracting from his designs, the totalitarian regime where no original opinion went unpunished would seem to be the last place to seek work. The Soviet Union under Nikita Khruschev had been embarrassed when English design expert Paul Reilly critiqued Russian design and found it wanting. Loewy was able to bring in a commission from the experience; in a March 19, 1973, memo to Bill Snaith, he outlines the contract as offering $300,000 per year for two years (more than $1.6 million per year today). Russian designer Yuri B. Soloviev suggested to Soviet bureaucrats that the state create a Soviet standard of design. To conceive a design standard and eventually form a state design institute, Soloviev brought in several major designers to consult.

"I did not know much about foreign design, but I knew the name Loewy," he wrote. Loewy was invited by the Committee for Science and Technology to visit the country and present lectures (Soloviev also later invited graphic designer Herbert Pinzke, a Chicago-based designer known for his work in typefaces, and industrial designer Samuel Sherr). Loewy came to Moscow in 1961 with Viola, and Soloviev escorted the couple throughout Russia, Georgia, and other states. Loewy tried to be unobtrusive but, as Soloviev tells it, the great designer was not well versed in idle conversation: "I was amusing my guests with various anecdotes and happened to mention that I was fond of aquatic sports, that I had a motorboat and enjoyed water-skiing. Loewy remarked that they also loved such activities, but had been compelled to sell their yacht to the king of Saudi Arabia. When questioned why, they explained they were unable to assemble a crew that was up to their standards. I couldn't help but laugh at the impossibility of small talk."

In a speech to the Harvard Business School, Loewy had no qualms about which consumer market was superior. "The citizens of Lower

Slobovia may not give a hoot for freedom of speech, but how they fall for a gleaming Frigidaire, a streamlined bus or a coffee percolator." He later characterized the Soviet Union, where he made ten visits between 1962 and 1976, as a developing country that had little idea about cost or how to pay a consultant. Dealing with the byzantine Soviet bureaucracy was frustrating as well, as various officials would appear to change specifications or plans in the middle of projects.

Loewy brought reams of design examples to Russia and made presentations, and the central bureaucracy decided to found a national institute for scientific research in technical aesthetics. Soloviev wrote, "Not only was Loewy a talented designer, but a brilliant propagandist of design, and in this capacity no borders existed or political differences existed for him. Moreover, he arrived at just the right time, at the height of the thaw. In my opinion, such a revolutionary government resolution would have been impossible two years later, in the period of stagnation under Leonid Brezhnev." Loewy would return to Russia many times and eventually proposed to Soloviev that his firm cooperate with the Russian design institute. He suggested a joint design firm with his French office, CEI. At their next meeting, Loewy was forced to renege on the agreement because several clients had threatened to withdraw their business over his work for the Soviet Union. Most Western clients feared the Soviets would use the collaboration as technical espionage.

Loewy enjoyed his relationship with the Soviet Union, but it was not without its pitfalls. "When the design is halfway completed they call you up and say 'no, we've decided to lengthen the wheelbase.' What seems to them a small change, affects, as you know, the whole concept of the integral. They don't understand that the designer has to start all over again. And they do not understand that this requires additional compensation since they tend to think of you as an employee rather than a free-lance designer, even when you are also a retained consultant."

As Soloviev's institute prospered, the designers found that many Soviet manufacturers were not using their designs. Soloviev suggested to Nikolaevich Smelyakov, an official of the Ministry of Foreign Trade, that the institute develop new products by enlisting a famous foreign designer: Raymond Loewy. The Soviets trumpeted the agreement as "the first USSR experiment in international cooperation in this field." Loewy was offered a contract worth close to $1 million to develop thirteen products: among them, a car (the Moskvich 412), a locomotive, a refrigerator (ZIL), and the Zenith camera. Of the projects accepted by Loewy, not one was produced. A dozen years after Loewy delivered the designs, two products were made, the Moskvich automobile and the ZIL refrigerator. Loewy's sketches for the Moskvich look sleek, anticipating the rakish lines of the Jensen Interceptor, the Ford Taurus, and the DeLorean. The completed final design, created by Moscow manufacturers, was somewhat less successful. It looked more like a 1968 Ford Falcon.

Loewy worked closely with David Butler, his vice president in the New York office, and kept the Moskvich commission from overstepping its bounds. Loewy urged that the firm provide only design recommendations and refrain from making technical suggestions. Loewy underlined this in a note to Butler. "Dave: From now on you are taking the responsibilities of an entire body division. Even at Studebaker where I had forty people on staff we never got involved to such a degree. We would make color renderings and some line drawings and send them to the body engineering division who would build mockups of the designs selected. We would only keep watching so they followed our ideas closely. Thirty or forty people were working on the jobs. Remember we are selling a design concept. We are not an automobile factory."

By the end of the Soviet experiment, Loewy's own business operations were on the brink of toppling into insolvency. Loewy's advanced age, coupled with his lax financial management style, had made for lean times in all three design offices. Loewy had always delegated financial management to in-house specialists, and this model worked well in his salad days, with money pouring in from new commissions. But as commissions slowed, the New York office had taken on debt. The hammer blow of Bill Snaith's early death during heart surgery was followed by a terrible recession brought on by the oil crisis of the 1970s. In 1973, Loewy employed 190 people in New York, 48 in Paris, and 20 in London. In 1975, a year after Snaith's death, Loewy merged all three corporate entities into one, titled Raymond Loewy International. Less than a year later, Loewy and his wife sold their shares in the business. Viola Loewy, explaining the downsizing of their lifestyle, told a reporter who asked how many homes the couple maintained: "Of course we have very little of that left. You can't get help—so you concentrate." In 1977, Raymond Loewy International declared bankruptcy. Most of the employees started their own firms, retired, or went to work for competitors.

The Loewys returned to France, where they continued to face financial strains. He sold lithographs of his most famous designs and also published *Industrial Design: Raymond Loewy* in 1979.

AFTER LOEWY'S DEATH IN 1986, Viola Loewy parceled out the legacy of his work when she decided to sell the designer's papers. The Library of Congress, which had accepted a 1981 donation of the archives of designers Charles and Ray Eames, had made it known that they would accept the entire archive. Unfortunately, Loewy's widow had to sell the archives to live. Most of the papers from Loewy's early career had been lost, so these lots primarily represented designs from his later work. The publishers of *ID Magazine*, Randolph McAusland and James F. Fulton, heard that the papers were to be auctioned off in France in Rambouillet. The publishers collected $30,000 from a variety of sources to buy archives for the Library of Congress. Fulton, who was a former manager of CEI, attended the auction with CEI director Evert Endt and purchased 33 lots of the collection. There were 793 lots of Loewy material at the auction, with the library donation representing just a fraction of the whole, including significant portions of his remaining art collection and furniture from his home in France. A Parisian art dealer, Jean Pierre Blusson, bought 30 lots. The rest of the estate went to individual bidders.

14

Legacy

The subtitle of the German edition of *Never Leave Well Enough Alone* translated to "The Experiences of One of the Most Successful Designers of Our Time." And indeed he was. Loewy assimilated into American culture almost from the moment he stepped ashore from the SS *France*. The first lesson he absorbed was that any person can be re-invented and attain success. His influence and reputation waxed and waned over the period of his lifetime, but his carefully nurtured legacy of superlative design created in service of "the rising sales graph" slowly came full circle.

Many of his contemporaries have faded into memory while Loewy's reputation has steadily regained its resonance as journalists and critics celebrate his mission to forego theoretical design ideology in favor of the rising sales curve. As trends and styles changed in the years after his death, the bold self-marketing that earned Loewy a reputation as a publicity hound was no longer scorned by tastemakers. Now it is called "branding." But regaining his reputation did not happen overnight.

True to his credo, Loewy never produced junky products, even as his career wound down. He also never diluted his image by merchandis-ing his name through myriad consumer products. That is not to say he wouldn't have followed the same path as modern-day "name" designers like Donna Karan, Calvin Klein, and style marketer Martha Stewart by creating epic product lines, but rather that those types of opportunities were much scarcer in Loewy's lifetime. Through astute publicity and an insistence on remaining the public face of his company, right down to every signature on design sketches, Loewy branded himself decades before the noun officially became a verb.

His company proved to be a launching pad for the careers of others, but as the design business changed, the Loewy business model was barely keeping up. At one time, Bill Snaith might have transitioned the company into a retail design firm, but his early death kept Raymond Loewy in charge, and nothing about Loewy's decisions in later years suggests he would have turned over the company to the extremely capable Snaith. As the profession morphed into increasingly specialized areas, the Loewy firm lost more clients, requiring the renowned designer to shed assets and downsize his operation.

After disbanding his firm in North America in the 1970s, Loewy made his London and Paris offices the focal point for his remaining design activity. Loewy did anticipate the global marketplace and fought to make his products the most advanced, yet acceptable to consumers beyond the parochial shores of the United States. Loewy acknowledged that American companies had to look outside their traditional consumers, saying designers "must live internationally to respect and accept the marketplace."

Loewy was prescient in seeking global clients, although acceptance was grudging at first. In France, in 1952, Loewy founded CEI. His reputation as a designer was largely gained in America; his entrance into France was met with resounding indifference. The French manufacturing ethic was not disposed to aggressive marketing, and the design ethos for Europe leaned more toward the purist Bauhaus model. The establishment of the Paris office emerged from a social encounter in Saint-Tropez between Loewy and the owner of the department store Bazar de l'Hotel de Ville, at which the designer was asked to redesign the store's trucks and sales kiosks used for industrial shows. After seeing CEI's initial designs, the commission was expanded to include redesigning stores. Using its hard-won expertise in retail design from Snaith's US operation, CEI gave European stores American-style interiors: bright, open, and less dependent on employee service.

For the first decade of CEI's existence, commissions came in the form of interior designs for banks, department stores, and Air France. Eventually, the firm expanded into product design, using a more minimalist style than the American home office. Loewy imported an old-fashioned American publicity stunt to open the office. He sailed a large boat that

had been dazzlingly refitted in England into the Quai d'Orsay and invited the cream of the Paris business community aboard to chat with
employees and inspect the firm's latest work. Some influential designs to
emerge from the French office included the Coquelle cookware line for
the French firm Le Creuset and the Elna Lotus sewing machine, a 1964
commission that modernized the look of a homely appliance into a sleek,
functional machine that remarkably resembles the twenty-first-century
machines that home seamstresses currently use. The Elna design was
chosen for the collection of the Museum of Modern Art. One of the firm's
first product triumphs centered on the cast iron cookware for Le Creuset,
which used an old-fashioned material in startling forms. The most advanced look was a sharply angled stewpot with a sleek nautical shape.

The CEI commission for Air France—interiors for the Concorde
supersonic jetliner—was Loewy's last major transportation project. The
company designed the jet's lighting, seats, and décor, and it also created
its minimalist flatware, china, and glassware. The cramped space of the
cabin made for some unique design problems. The seats had to be outfitted with headrests because of the high-speed takeoffs. The company
covered the seats with various colored fabrics that did not repeat in the
seating pattern. Loewy's reasoning was to keep the passengers happy,
preferably unaware they were flying through the air in a pencil-thin
tube. The narrow width of the fuselage forced Loewy and the designers
to create an optical illusion of width and height by placing a jet-black
band down the center of the interior ceiling that made the central aisle
seem wider than it was.

Within a few years, CEI was tackling huge projects, including a
worldwide redesign of British Petroleum service stations. In 1960, Loewy
appointed Douglas Kelley as head of the office, but the French staff soon
rebelled against his "American" management style, and Loewy allowed
the shop to function in the European mode of relaxed work schedules,
more vacation time, and less hierarchy. CEI became much more independent from Loewy's American office, eventually competing with the
"family" firms for business. The shop won the 1967 contract for Shell
International, for example, which at the time was the largest design contract ever awarded to a European design agency. The campaign ranged
from station architecture to service uniforms to packaging. It also re-
vamped gas pump design. The commission, which lasted until 1976,
energized Loewy as well. He organized research trips to study how consumers in Japan, Italy, and Canada used Shell products.

As the European offices shone more brightly, in 1975 Loewy decided to "semi-retire" and sold off the client accounts for the New York office, but he retained the London and Paris operations, establishing himself as a paid consultant to both. His own image dimmed, Loewy decided in 1980, at age 87, to fully retire. He sold the company name to the London office to Farrell and partner Thomas Riedel, retaining the rights to the French operation. Three years later, Loewy sold the shares of the Paris branch to Farrell.

The designers on the Paris staff did not have a chance to take over the firm. Instead, the aging designer, in need of a nest egg for his retirement, felt he could get a higher price from outside investors. The breakup of the company was Europe's gain, as a series of CEI staff design professionals established boutique design shops, many of which still exist.

Retirement did not mean Loewy was going to be ignored. His genius for branding himself was not without a common touch, sometimes leavening the hard sell with a touch of self-deprecating humor. In a speech to the Society of Automotive Engineers, Loewy poked genteel fun at his image:

> It is generally accepted that an industrial designer works best when draped at the edge of a turquoise swimming pool while Nubian slaves in gold sarongs serve chilled nectars in silver cups. Soft music relaxes his nerves while a blonde masseuse works on the master's right wrist. His wrist is a little stiff from having endorsed too many checks with a ballpoint pen. After being rubbed with pine oil, the industrial designer is ready for the morning chores, *consisting of interviews with the nation's foremost magazine writers in search of new material.* He then makes ready for the afternoon session, which is devoted to press photographers. And so on and so on.

By the end of his career, with no Nubian attendants in sight, Loewy could only adapt to the changing environment on his own terms. As he aged, he could not break loose of the methods that had served him so well in keeping his name in the public eye. "Loewy, who had transformed American icons, ended up crystallized as if in amber, caught in his own trap, transformed into an over-designed product himself," wrote Philippe Tretiak.

Design historian Stephen Bayley diagnoses—a bit harshly—Loewy in his book *Design: Intelligence Made Visible* as having a "narcissistic personality disorder defined as a pervasive pattern of grandiosity, need for admiration and lack of empathy." Yes, he was self-involved, aloof, and put his name on everything ever to emerge from his office, but there's little evidence in others' recollections that he was an overbearing egomaniac. After all, Loewy raised a well-adjusted daughter who became a respected journalist, and his former employees rarely went on the record about his perceived shortcomings. Most of the designers who wrote memoirs or remembrances of Loewy forgave his self-centered persona and praised his willingness to give them varied experiences and to offer his blessing if they chose to go out on their own.

Although reams of words were generated in praise of Loewy as a genius in French cuffs, his image was finally cemented in public memory over the years through publicity photos. This studied photographic legacy, inspired by the movie studio glamour shots epitomized by the work of George Hurrell and others, was a precursor to the celebrity journalism pioneered by *People*, *US Weekly*, and other gossip magazines. Most of these images show Loewy leaning against a product he designed, be it a locomotive, helicopter, or automobile. He had no qualms about how cool or ridiculous he looked. One photograph of the designer and his Avanti, taken at his Palm Springs home, shows the courtly designer outfitted in western garb. He has on jeans with 3-inch rolled-up cuffs and a yoked western shirt. The pose is one of utter relaxation. Another one, which ran in *Life*, shows the designer and Viola near their dune buggy. He's wearing a casual sweater and slacks, complemented by large goggles. Viola Loewy is in a western outfit of her own: skintight pants, knee-high boots, and a long, fringed vest, set off by a beaded headband. Design historian Jeffrey Meikle lays out the effect these photographs had: "Projecting his personal image of the industrial designer, Loewy posed as the heroic individual, the man with impeccable elegance and flair, in touch with the spirit of the time, capable of creating forms attractive to the American people while maintaining the clinical detachment to inspire confidence in the toughest of business clients."

Thirty years after the creation of his "heroic designer" façade, the dimming of Loewy's reputation began in the 1960s and continued until his death. After the initial deification of industrial designers in the 1920s and '30s, intellectuals and art critics writing in art and aesthetics journals subscribed to the same theories as the fine art world—that "pure" design and minimalism were the ideals to strive for. This trend intensified when Bauhaus-influenced immigrants such as Gropius and Moholy-Nagy were hired to teach at Harvard and the Illinois Institute of Technology, respectively. Their lessons would foment critical thinking on design and aesthetics for the next thirty years. A series of exhibitions at New York's Museum of Modern Art cemented the reputations of Bauhaus-influenced designers and relegated the pioneering designers such as Dreyfuss, Teague, and Loewy to the status of "stylists."

As Loewy left the field of industrial design in the late 1970s, the movement of designers into corporate positions had all but become policy. There were still design boutiques, a few on the scale of Loewy's operation, but it became extremely difficult for one design shop to be all things to all clients.

Loewy's elder statesman status occasionally commanded respect from journalists seeking a quote on the latest design or from curators looking for reactions to an exhibit. Despite the low opinions his work garnered from art critics, some museum curators realized the importance of Loewy's career and began asking him to participate in exhibitions.

The reconsideration of Loewy's career began in earnest with an exhibition of his designs at the Renwick Gallery, the design museum for the Smithsonian Institution in Washington, DC. The exhibit began with photos of the Gestetner duplicator and segued into renderings of the Skylab proposals, the 1953 Studebaker Starliner coupe, the S-1 locomotive for the Pennsylvania Railroad, the Greyhound Scenicruiser, and the interior and exterior of *Air Force One*. The exhibition was immediately popular, but not everybody was a fan. "I think one real blow to Loewy's popularity with his former employees was that terrible exhibition of his in 1975," recalled designer John Ebstein. "He had collected his employees' old renderings, signed them with his own name—I saw some of my drawings up there with his signature—and then offered them for sale. But this was after his office had closed. He was getting old and scared: he wanted to make sure his name and fame remained in the public eye. It was quite sad but I know it was deeply resented by many designers."

Laurence Loewy, writing on raymondloewy.com, recalled that her father called her at the University of Southern California, where she was enrolled, to say, "Laurence I'm thrilled, the Smithsonian Institution is doing a retrospective of my career. It is a tremendous honor." Laurence said Loewy considered the exhibition to be his ultimate tribute. In the Loewy archives at the Hagley Museum and Library is an eighteen-page speech that Loewy delivered at its opening. "We are the leaders of a modern Crusade against any manufacturers who will further burden the already harassed consumer with added grief, irritation, and waste."

In his introduction to the catalog for the exhibit, Joshua Taylor, director of the National Collection of Fine Arts, presented as accurate a description of Loewy's influence as any biographer or design critic. "Raymond Loewy has been a great creator of public symbols. A hard-headed technician, he also deals in myth, the kind of myth by which society lives. The scientific knowledge and procedure on which his work is based seem always to have been at the service of a goading dream of a fascinating future world. It is a world in which cars go faster, mechanisms are eager to function well for the good of man, and every form slips easily into its purposeful role."

Shortly thereafter, in 1976, Loewy began work on *Industrial Design* at his home in Palm Springs. Laurence Loewy wrote on the Raymond Loewy website:

> Dad had just closed his New York office and felt the need to document his most significant designs for future generations. As an aspiring journalist enrolled at USC, I vividly remember stepping into the studio when Dad was compiling photos and writing the footnotes for each illustration. His small studio looked like a bomb hit it. There were hundreds of photos and texts organized in rows lining every square inch of the beige carpet. In the center of the workplace, stood a desk with Dad hovering before it, ruler in one hand and a pencil in the other. He labored tirelessly over the project; sometimes resuming at dawn and working through dinner. Dad's image was at stake and he knew it.

She continued describing her father's efforts to set down his accomplishments on paper, noting his awareness that his time had passed. "During the 1970s U.S. industrial design was no longer in vogue. The majority of Dad's clients had established in-house design teams. Loewy, the sole surviving member of the industrial design founders; Dreyfuss, Bel Geddes and Teague, felt somewhat disillusioned in the U.S."

The family's decision to sell the Loewy archives, exacerbated by business reversals associated with the company's design commissions in the Soviet Union, highlighted the need to establish collections of top American design in the nation's museums and archives. The Cooper Hewitt National Design Museum took the archives of Deskey, Dreyfuss, and other designers, but few museums had the funding—or, in some cases, the inclination—to archive the work of the nation's greatest designers. The collection of Charles and Ray Eames's work was taken in by the Library of Congress as well as by Germany's Vitra Design Museum, which holds work by George Nelson and other American designers. A February 17, 2000, story in the *New York Times* shone a light on the dearth of comprehensive design collections. Writer Julie Iovine cited the case of Minoru Yamasaki, the architect of the World Trade Center, whose papers were destroyed by his widow when no museum or archive accepted them. Contemporary designers and architects, such as Michael Graves and Frank Gehry, have a different problem. Prolific in the extreme, many designers and architects are personally storing their work to preserve it for a possible sale or donation. Iovine used the example that Gehry's archived designs and artifacts for a single project, an unbuilt house for billionaire Peter B. Lewis, takes up 200,000 square feet of space.

The auction of Loewy's archive netted about $500,000 for Loewy's family and scattered the rest of the original collection of papers among various collectors. At the time of his death, Loewy had not yet regained the renown he once enjoyed. That journey would take another decade or so. Along the way, didacticism gave way to eclecticism in art criticism, and more writers celebrated the accomplishments of crowd-pleasing artists. Loewy—along with artists working with other media, such as Andrew Wyeth, Norman Rockwell, Douglas Sirk, and Louis Armstrong, to name a few—found a new audience among consumers who began to purchase affordable design. Products such as Morison Cousins's Tupperware, Jack Telnack's Ford cars, Michael Graves's housewares, and even Jonathan Ive's original iMac all subscribed to Loewy's idea that, given a choice between two similar products, the better-looking item will sell more effectively.

Time magazine once again was instrumental in resurrecting Loewy's name. *Time* art critic Robert Hughes mentioned Loewy's influence in many reviews, and the magazine used a Raymond Loewy quote about rising sales curves as a lodestone for a 2001 article on design.

A CASE CAN BE MADE that Loewy is, as he consistently claimed, the father of industrial design, but a far more influential contribution was using his considerable accomplishment as a businessman to create a template of success for others to follow. Ideally, Loewy could have used his marketing skills to promote himself as the founding father of his profession. After all, he outlived his nearest rivals by decades and worked until about ten years before his death. That longevity and facility for promotion worked against him initially, as the art world and media critics railed against the "styling" and "decorative" skills of Loewy and others to support the minimalist aesthetic of the International and Bauhaus tradition. Critics, in general, prefer that artists be grateful to be anointed, rather than anoint themselves through hard work and shrewd promotion.

Loewy's assimilation of Betty Reese's lessons in public relations created a schism between design and art tastemakers who, over time—particularly as tastes changed in the latter half of the twentieth century—saw Loewy as too blatant a booster for his own work. The many homes, myriad cars, and constant presence in the press made the critics of the time weary of hearing his name.

Oddly, especially for a man who carefully planned every meeting and presentation, the lack of planning for a business successor virtually ensured that his name and business would not survive him. His name does live on as a business, as do Henry Dreyfuss and Associates (today based in Ann Arbor, Michigan) and Brooks Stevens Design and Deskey Associates today. These firms continue to create designs for a variety of products. Unfortunately, the Loewy business name survives only as Loewy, a European marketing company, which had eliminated product design from its services until 2007, when it merged with Seymourpowell, a design and marketing company. That his design studio is no longer operating in the United States, where Loewy found his purpose and immeasurably contributed to a new American way of life, is one of the misfortunes of a long life lived well.

Raymond Loewy's ultimate legacy is not only his principles of design but also the creation of a brand personality. As the man who could find the camera in any room he walked into, Loewy was able to sell his work through the sheer force of personality. The presentation of image before substance would never leave him, even when it was clear that revealing a substantial talent would better serve his image. His penchant for publicity, whether it was inserting himself into society photographs or participating in thrill sports—diving, racing—never ebbed. In 1963, he enrolled in a high-performance driving course with racing legend Carroll Shelby at the Riverside racetrack outside of Los Angeles. He was 70 at the time. Even in retirement, his image remained all-important. "A flashy dresser, he often sports window-pane-checked jackets with contrasting striped shirts and polka-dot socks," wrote Susan Helen Anderson in a 1979 *New York Times Magazine* profile written seven years before Loewy's death.

The public man was seen as a bon vivant and tastemaker, and Loewy worked hard at burnishing his social skills. With his wife Viola, he was unparalleled at creating the image of success. Unlike his friend Henry Dreyfuss, there was no chance Raymond Loewy would ever take a meeting wearing a plain brown suit. Yet behind the silk cravats, French cuffs, and brightly patterned hose, Loewy was much different than his public persona. His hobbies were pastimes performed alone: deep-sea diving and racing. The public Raymond Loewy also published two of the most readable books on industrial design. *Never Leave Well Enough Alone* remains entertaining fifty years later and, as noted previously, it remains the only book written by a major designer that truly reveals, hidden beneath Loewy's "schmaltz," how to run a business. *Industrial Design* is less entertaining, but it gives succinct portraits of Loewy's career achievements in a colorful and individual style. By comparison, the books by Henry Dreyfuss, Norman Bel Geddes, Walter Teague, and Harold Van Doren read like treatises or textbooks. He may have been self-involved, but it's hard to imagine Loewy publishing a 200-page lecture on aesthetics. While not classic literature, *Industrial Design* still functions as the best purely visual introduction to design principles, and *Never Leave Well Enough Alone* offers a jocular look at the life of an "honorary" American businessman. Although dated, it still offers a good summary of stylish living.

The world described in his writing is one of discovery. Loewy is continually onto the next new thing. He moves to the next sale, the next product. His world is always written in the future tense; there is no looking back, and what is on the page is always the sunniest version of the story. His philosophy was summarized concisely during an interview: "I believe in natural talent; if you don't have it, do something else. Don't bother about sociology and ideology, what you need is a good knowledge of engineering, paper, pencils and a slide rule, common sense and respect for the arts of the past."

He loved to remind people of his long-lasting career. Moreover, he loved retelling his past triumphs. On his desk in his Paris home was an acrylic plastic bar in which was embedded two Lucky Strike packages, a mobile before-and-after example that showed the Loewy redesign next to the original packaging. It functioned as equal parts art and advertisement—a portable branding statement.

If Loewy's pronouncements on bad design, whether decrying the spinach and schmaltz of GM products or the overdone typography of a lunchmeat package, were meant to poke other designers to greatness, the end result was usually resentment or retaliation. In part because many designers had an unspoken aversion to Loewy's cultivation of his "exotic European" persona, few targets saw his comments as constructive criticism. Bill Mitchell, GM's successor to Harley Earl, was often quoted taking shots at "refrigerator designers" and "European-influenced" designs. Those who worked for Loewy seemed to forgive his faults and his willingness to throw darts at his competition and instead recognized his generous offers of opportunity.

Even the insular world of car design was liberated by Loewy's ability to allow designers to switch specialties. Studebaker designer Bob Bourke remembered his good fortune at working on many different accounts in the lean years of the Studebaker account and after Loewy's company and the car manufacturer parted ways. Gordon Buehrig, Bob Andrews, and other Loewy car designers had similar stories.

Most of Loewy's coworkers and rivals agree that he was a legitimate pioneer in his chosen profession. Friends and competitors all saw that Loewy built an empire from building accounts through vision, salesmanship, and delivering, as Studebaker's motto went, "more than what was promised."

Many design historians have a love/hate relationship with the Loewy legacy, admiring his innate feel for American consumer taste and breadth of accomplishment, but deriding his assiduous courting of the media. Interestingly, such "marketing" is forgiven in artists who emerged later, such as John Lennon (who cannily cultivated rock journalists who obligingly elevated Lennon over the more musically talented Paul McCartney); screenwriter Robert Towne (who, with director Robert Altman, made a point of flattering critic Pauline Kael); and abstract expressionist painters (who produced works hewing to the theories of critic Clement Greenberg).

Loewy's personal involvement in his firm's design work also is a sensitive issue for art and design historians. Loewy recognized early that his talents were not on the drafting board. Aside from his fashion illustration, most of his early designs were executed with the help of modelers hired first as freelancers. Even the famous Gestetner commission was executed by a clay modeler working under Loewy's discerning eye. Nearly all the designers who created large agencies assumed roles as editors and supervisors early on. Walter Teague famously became estranged from his son after taking credit for his son's design of the Marmon 16 automobile. Henry Dreyfuss, perhaps the most talented designer on the drawing table, handed off the bulk of his agency's work to staff. Certainly, car designers such as Harley Earl, who was also a poor draftsman, and Bill Mitchell, who was a fine artist, could not have personally designed the massive output of GM models. Instead, they were exceptional editors, albeit with a uniquely cracked methodology—Earl by prowling studios after hours and destroying models he didn't like, and Mitchell by profanely disparaging designs he found lacking. Those management methods make Loewy's credit-grabbing seem almost reasonable.

Tim Brown, president and CEO of the design firm IDEO, published an essay in *Metropolis Magazine* adapted from his book *Change by Design: How Design Thinking Transforms Organizations and Inspires Innovation* that deftly captures the reconsidered image of Loewy, Dreyfuss, and their contemporaries. He wrote:

> What they all shared was optimism, an openness to experimentation, a love of storytelling, a need to collaborate, and an instinct to think with their hands—to build to prototype, and to communicate complex ideas with masterful simplicity. They don't just do design; they lived design. These great thinkers were not as they appear in the coffee table books about "pioneers," "masters," and "icons" of modern design. They were not minimalist, esoteric members of design's elite priesthood, and they did not wear black turtlenecks. They were creative innovators who bridged the chasm between thinking and doing because they were passionately committed to the goal of a better life and a better world.

Loewy could not have phrased it better himself.

Unlike many men who reach an advanced age, Loewy never seemed to pine for the past, and he never lost his enthusiasms. He was interviewed by Morley Safer for a *60 Minutes* profile in 1979 and delivered a courtly and entertaining interview. Amid a synopsis of his long career, Loewy and Safer find time to tour a hardware store, where

Loewy—resplendent in ascot, sports jacket, and sunglasses—critiques the design of everything from mousetraps to toilet seats.

ULTIMATELY, LOEWY'S INCESSANT selling of himself, of his design firm, and of industrial design as a profession—a personality trait annoying to his contemporaries and to most automotive and design historians— became the gold standard for marketers. Loewy and Betty Reese instinctively understood that marketers who courted the media solidified their reputation with the only critics that mattered: the consumer public.

In Loewy's business lifetime, from the 1930s to the 1970s, aggressive advertising was perceived as being somehow gauche, and pursuing publicity was distasteful to most service professions. Within a decade of Loewy's retirement, doctors, lawyers, and bankers were elbowing each other aside to get on television, radio, and in print. By the 1990s, designers, architects, and business executives established themselves and their reputation not by slow and steady word of mouth, but through publicity. Loewy's contemporaries excoriated him for his brazen boasting and bald-faced publicity moves. Raymond Loewy was once again decades ahead of his time.

Oddly, Loewy's work for NASA, designs he felt would define the final phase of his career, has turned out to be his most forgettable. In one sense, the Skylab work was destined to oblivion as NASA moved away from orbiting living habitats into the space shuttle program. Still, given that most of the designs created for the Skylab mission were generic "living environment" ideas, nothing in the reams of sketches for the project rises to the level of immortality and are just barely memorable. Loewy, who waxed so enthusiastically about demanding the astronauts have a porthole, sort of missed the house needed for the window. In the late 1970s, Loewy couldn't have known the space program would be de-emphasized, effectively eliminating the need for handsomely designed living spaces. He thought space was the next great stage for design and focused most of his remaining public cachet on his work there. The cover of *Industrial Design*, as noted previously, shows Loewy posing with his NASA work, although most people would not instantly recognize what they were looking at. On the back cover, a much more relaxed Loewy poses, cigarette in hand, sitting among models and samples of his work—a Greyhound Scenicruiser, a six-pack of Coke bottles, a Rosenthal teacup, a Shell oil can—all more recognizable to the twenty-first-century consumer. Always able to see where the marketplace was headed before taking on the NASA work, Loewy misjudged the eventual legacy of the space program. He had hitched his final years to a waning star.

If journalists or essayists need a pithy historical quote on design, they turn to Loewy clippings. The absolutist culture of the 1950–80 art world has long since dissipated to an atmosphere where most critics are more accepting of multitasking, multitalented artists working in a variety of media. Today's opinion makers also value savvy marketers who know how to "spin" reporters, critics, or even gossip columnists. Designers who boldly create personas for their brand and seek out publicity for that brand are celebrated as business geniuses. What are *Martha Stewart Living* or *O* but magazine versions of *Never Leave Well Enough Alone*? Loewy has become a touchstone for a new generation of designers. To some extent, the Internet has facilitated Loewy's rising reputation in recent years. His daughter Laurence played a central role in his renaissance by creating and curating raymondloewy.com. She also created Loewy Design LLC to promote her father's work and served on the board of the Raymond Loewy Foundation from 2000 to 2005. She collaborated with Glenn Porter, curator of the Hagley Museum and Library to create the exhibition *Raymond Loewy: Designs for a Consumer Culture*. Laurence, who died on October 20, 2008, had been working on a documentary film and several books about her father. Today, Laurence Loewy's husband, David Hagerman, carries on her work by managing the Loewy estate. The future of the Raymond Loewy Museum of Industrial Design, restoration and cataloging of artifacts, site planning, and architecture, rests with David Hagerman, museum CEO, and the Raymond Loewy Museum board of directors.

UPON HIS DEATH, *ID Magazine* celebrated Loewy as the "Last of the Magicians." "There are not many men whose dying genuinely marks the end of an era, but Loewy was one of them. For he *was* the last of the magicians, the last designer credited with transmogrifying products through a mysteriously intuitive process suspiciously reminiscent of the laying-on of hands." Loewy never revealed much about his creative process. He came closest in the book *Industrial Design*, writing, "I believe one should design for the advantages of the largest mass of people, first and always. . . . I think one should try to elevate the aesthetic level of society . . . Design simplicity is the essence; the main goal is not to complicate further the already difficult life of the consumer. Our goal is to give him some peace of mind. Junky stuff is consumer murder."

Notes

Introduction

p. 7: **"All you want is to get personal publicity"**: Notes written by Raymond Loewy, Dec. 11, 1965. Raymond Loewy Archive, Hagley Museum and Library.

p. 8: **"I realize American Motors"**: Letter to Robert Beverly Evans, Nov. 2, 1966. Raymond Loewy Archive, Hagley Museum and Library.

Chapter 1. New Shores

Books and articles used to flesh out Raymond Loewy's early life include his auto-biography, *Never Leave Well Enough Alone*, and *Raymond Loewy* by Paul Jodard. The background information on turn-of-the-century Paris and France and Paris during Raymond Loewy's formative years comes from *A History of Modern France* by Jeremy Popkin, *Modern France, 1880–2002* edited by James McMillan, and *Seven Ages of Paris* by Alistair Horne.

p. 10: **"A young man who came to America"**: Loewy, *Never Leave Well Enough Alone*, viii.

p. 11: **"Clarity was not always a strong point"**: Jodard, *Raymond Loewy*, 14.

p. 12: **"It is better to be envied than pitied"**: Albin Krebs, Raymond Loewy obituary, *New York Times*, July 15, 1986.

p. 13: **"The only country with a 300-metre flagpole"**: Horne, *Seven Ages of Paris*, 326.

p. 14: **observers called the structure**: Horne, *Seven Ages of Paris*, 326.

p. 14: **Chastenet called Paris "a work of art"**: Horne, *Seven Ages of Paris*, 289.

p. 15: **such as his father's conversion to vegetarianism**: see Loewy, *Never Leave Well Enough Alone*, 19.

p. 15: **Unable to resist showing off his business acumen**: Loewy, *Never Leave Well Enough Alone*, 21.

p. 16: **[he] filled his school notebooks with images of trains and automobiles**: "Up From the Egg," *Time*, Oct. 31, 1949.

p. 16: **"My greatest thrill was to feel with my hand the honeycomb"**: Loewy, *Never Leave Well Enough Alone*, 26–27.

p. 16: **"They effectively influenced me"**: Loewy, *Never Leave Well Enough Alone*, 26–27.

p. 18: **"It helped me overcome a great shyness"**: Loewy, *Never Leave Well Enough Alone*, 29.

p. 19: **Raymond received grades that reflected the tunnel vision**: Loewy, *Never Leave Well Enough Alone*, 44.

Chapter 2. Portrait of the Young Engineer as an Artist

Never Leave Well Enough Alone provided many of the first-person impressions of Loewy's early years in America. Other invaluable sources for his arrival and initial forays into business are *Industrial Design*, Loewy's 1979 collection of career reminiscences; *Raymond Loewy and Streamlined Design* by Philippe Tretiak; *Raymond Loewy, Pioneer of Industrial Design* edited by Angela Schoenberger; and *Raymond Loewy* by Paul Jodard. Information on the rise of industrial design as a profession and the men who started industrial design as a profession came from such books as *Twentieth Century Style and Design* by Stephen Bayley, Phillipe Garner, and Deyan Sudjic; *Twentieth Century Limited* by Jeffrey Meikle; *The Streamlined Decade* by Donald Bush (p. 16); *Henry Dreyfuss, Industrial Designer* by Russell Flinchum; and *Designing for People* by Henry Dreyfuss. Source material for the interaction of advertising, the American marketplace, and American consumers comes from *Domestic Revolutions* by Steven Mintz; *Land of Desire* by William R. Leach; *The Mirror Makers* by Stephen Fox; *Fables of Abundance* by Jackson Lears; *Advertising the American Dream* by Roland Marchand; *Twentieth-Century Style and Design* by Stephen Bayley, Phillipe Garner, and Deyan Sudjic; and *An All-Consuming Century* by Gary S. Cross.

p. 22: **"The giant scale of all things"**: Loewy, *Never Leave Well Enough Alone*, 10.

p. 22: **"He looked like a magnified version"**: Loewy, *Never Leave Well Enough Alone*, 57.

p. 22: **the new hire was not to be a "clock puncher"**: Loewy, *Never Leave Well Enough Alone*, 57.

p. 22: **"At the time, the technique was to [bring in] a truckload of stuff"**: Loewy, *Never Leave Well Enough Alone*, 57.

p. 22: **"It was dramatic, simple and potent—it sang"**: Loewy, *Never Leave Well Enough Alone*, 58.

p. 22: **"They were talking in hushed tones"**: Loewy, *Never Leave Well Enough Alone*, 58.

p. 23: **After his brief debacle at Macy's**: Loewy, *Never Leave Well Enough Alone*, 58.

p. 23: **"I imagined that a time would come"**: Loewy, *Never Leave Well Enough Alone*, 10.

p. 23: **refrigerators, toasters, vacuum cleaners, fans, stoves, and dishwashers**: Leach, *Land of Desire*, 270.

p. 23: **"I always abhorred the role of being a spectator"**: Loewy, *Industrial Design*, 10.

p. 24: **By 1929, Marion, Ohio**: Leach, *Land of Desire*, 274.

p. 24: **"There is no need to worry"**: Loewy, *Never Leave Well Enough Alone*, 33.

p. 24: **"You say one has to have personality"**: Evert Endt, "A Frenchman in New York," in Schoenberger, *Raymond Loewy*, 33.

p. 25: **"Financially, I was successful but I was intellectually frustrated"**: *New York Times*, Nov. 4, 1979.

p. 26: **Stores "crammed to the ceiling"**: Loewy, *Never Leave Well Enough Alone*, 74.

p. 26: **"My French friends can't believe, even in a small village"**: Loewy, *Never Leave Well Enough Alone*, 74.

p. 27: **"I felt I belonged here [in the United States]"**: Loewy, *Never Leave Well Enough Alone*, 77.

p. 28: **"looked like a very shy, unhappy machine"**: Loewy, *Never Leave Well Enough Alone*, 82.

p. 28: **"covered with a mysterious bluish down"**: Jodard, *Raymond Loewy*, 22.

p. 28: **"I often kidded Sigmund"**: Loewy, *Never Leave Well Enough Alone*, 61.

p. 30: **"He adopted the marketing philosophy of his country"**: Tretiak, *Raymond Loewy and Streamlined Design*, 8.

p. 30: **"The goal of design is to sell"**: Tretiak, *Raymond Loewy and Streamlined Design*, 8.

p. 31: **"He lives in an atmosphere of action"**: Lears, *Fables of Abundance*, 180.

p. 31: **"Consumer goods were the building blocks"**: Cross, *An All-Consuming Century*, 18.

p. 31: **"The car was the bellwether commodity of the new century"**: Cross, *An All-Consuming Century*, 25.

p. 32: The term **"industrial designer"** was first used in 1919: Meikle, *Twentieth Century Limited*, 40.

p. 33: **"The window is a stage, the merchandise as the players"**: Bush, *Streamlined Decade*, 50.

p. 34: **"The man who when asked to design a product"**: Bush, *Streamlined Decade*, 53.

p. 34: **"He seemed immersed in the establishment"**: Meikle, *Twentieth Century Limited*, 56.

p. 34: **"Make machines fit people—don't squeeze people into machines"**: Meikle, *Twentieth Century Limited*, 56.

p. 35: **"What we are working on is going to be ridden in"**: Flinchum, *Henry Dreyfuss*, 56.

p. 35: **"the only authentic genius this profession has produced"**: Flinchum, *Henry Dreyfuss*, 27.

p. 35: **"An honest job of design"**: Flinchum, *Henry Dreyfuss*, 27.

p. 35: **"If people are made safer, more comfortable"**: Dreyfuss, *Designing for People*, 24.

p. 36: **"Ours is the ever-changing battleground"**: Dreyfuss, *Designing for People*, 24.

p. 36: **"He was more interested in living than designing"**: Meikle, *Twentieth Century Limited*, 60.

p. 37: **By 1931, Young and Rubicam Agency had renamed**: Meikle, *Twentieth Century Limited*, 128.

p. 37: **"If we are to have beauty in the machine age"**: Meikle, *Twentieth Century Limited*, 130.

p. 38: **"I'm sure my lifestyle was an easy target for other designers"**: Loewy, *Industrial Design*, 18.

Chapter 3. The Artist (and Others) Shape the Things to Come

The material on the history and development of American industrial design is taken largely from the invaluable books *Twentieth Century Limited* by Jeffrey Meikle; *Designing Modern America* by Christopher Innes; and *American Design Ethic* by Arthur J. Pulos. The personal stories of designers at the beginning of the age of industrial design include *Never Leave Well Enough Alone*; *Horizons* by Norman Bel Geddes; and *Design This Day* by Walter Dorwin Teague.

p. 39: **"One of my daily morning task assignments"**: Budd Steinhilber, "Loewy Pencil Sharpener," Nov. 4, 2008, http://deconstructingproductdesign.com /loewy-pencil-sharpener/.

p. 41: **"Everything that moved through air or water"**: Meikle, *Twentieth Century Limited*, 148.

p. 41: **"The car was never built, owing to psychological factors"**: Bel Geddes, *Horizons*, 55.

p. 41: **"It will fly much more smoothly"**: Bel Geddes, *Horizons,* 111.

p. 41: **Christopher Innes calls the plane's design "certainly practical"**: Innes, *Designing Modern America*, 113.

p. 42: **"All the industrial design we have had in the United States"**: Bel Geddes, *Horizons,* 293.

p. 45: **American manufacturers and citizens responded with "a frenzy of inventions based upon gasoline engines, electric motors and heating coils"**: Pulos, *American Design Ethic*, 228.

p. 45: **"Automobiles were horseless carriages, parlor or cooking stoves were baroque idols"**: Pulos, *American Design Ethic*, 242.

p. 47: **"The mystery is that this particular man should have ever have designed anything at all"**: Gilbert Seldes, "The Long Road to Roxy's," *New Yorker*, February 25, 1933.

p. 47: **"Designers and manufacturers were always aware"**: Bayley and Conran, *Design*, 17.

p. 48: **"The collection has been brought to America"**: American Association of Museums, *Selected Collection of Objects*.

p. 48: **"The theater is a fickle mistress"**: Bel Geddes, Letter to the *New York Times*, Jan. 27, 1929.

p. 49: **"Industrial design emerged in the United States as a distinct calling"**: Pulos, *American Design Ethic*, 118.

p. 51: **Auto historian Beverly Rae Kimes called the Marmon a "beautifully proportioned, pace-setting design"**: "Antiques: A Car Design at Age 19 Set a Career," *New York Times*, June 15, 2001, E36.

p. 51: **"The entire car had a dynamic look of motion"**: Loewy, *Industrial Design*, 56.

p. 52: **"I was on the verge of being retained by the Chrysler Corporation"**: Loewy, *Never Leave Well Enough Alone*, 84.

p. 52: **"I believe it was the beginning of industrial design"**: Loewy, *Never Leave Well Enough Alone*, 84.

p. 52: **"I took the *Detroiter* [train] to Detroit, where a company driver waited for me"**: Loewy, *Never Leave Well Enough Alone*, 84.

p. 52: **"What he wished me to understand, see"**: Loewy, *Never Leave Well Enough Alone*, 84.

p. 53: **"It was a clear case of the polite brushoff"**: *Loewy, Never Leave Well Enough Alone*, 86.

p. 53: **"I knew that if they could see the design in the flesh"**: *Loewy, Never Leave Well Enough Alone*, 87.

p. 53: **"What was most objectionable was the fact"**: *Loewy, Never Leave Well Enough Alone*, 88.

p. 53: **"too high, too static and blunt-looking"**: *Loewy, Never Leave Well Enough Alone*, 89.

Chapter 4. Birth of a Salesman

The books that provided a critical picture of Raymond Loewy's early years as an industrial designer include *Never Leave Well Enough Alone* and *Designing for People* by Henry Dreyfuss. The invaluable information on the history of Sears, Roebuck & Company came from *Catalogs and Counters* by Boris Emmet and John E. Jueck and *Shaping an American Institution* by James C. Worthy. Critical reviews of Loewy's early designs that provided insight into how industrial design developed came from the books *Objects of Desire* by Adrian Forty; *The Total Package* by Thomas Hine; and *Twentieth Century Limited* by Jeffrey Meikle.

p. 56: **"The country was flooded with refrigerators"**: Loewy, *Never Leave Well Enough Alone*, 77.

p. 56: **"Competition would become fierce"**: Loewy, *Never Leave Well Enough Alone*, 77.

p. 57: **"I felt I belonged here"**: Loewy, *Never Leave Well Enough Alone*, 77.

p. 57: **"Some persons think the industrial designer is the equivalent of a wonder drug"**: Dreyfuss, *Designing for People*, 191.

p. 57: **"No one in the manufacturing world"**: Loewy, *Never Leave Well Enough Alone*, 115.

p. 57: **"Who is that fellow anyway, reeking with a foreign accent"**: Loewy, *Never Leave Well Enough Alone*, 115.

p. 58: **"And so on and so on, for weeks, for months, for years"**: Loewy, *Never Leave Well Enough Alone*, 117.

p. 60: **"When we started our design, the Coldspot"**: Loewy, *Never Leave Well Enough Alone*, 127.

p. 61: **"many models with wood casings did not sell the hygienic message"**: Forty, *Objects of Desire*, 179.

p. 61: **"So seamless [that] a spot called for instant removal"**: Forty, *Objects of Desire*, 180.

p. 62: **"He designed a product that people felt good about":** Hine, *Total Package*, 109.

p. 62: **"Form followed function if the function of a product is to be sold":** Hine, *Total Package*, 115.

p. 63: **the original Coldspot design was "a step in the evolution toward perfection"** Meikle, *Twentieth Century Limited*, 104.

p. 64: **"We had to contend with a group of twenty or thirty crackpot commercial artists":** Loewy, *Never Leave Well Enough Alone*, 128–29.

p. 64: **"So I took an office on the fifty-fourth floor":** Loewy, *Never Leave Well Enough Alone*, 91.

p. 65: **"She spoke fluent French":** Loewy, *Never Leave Well Enough Alone*, 99.

p. 65: **"Nothing is to come out of R. L. offices":** Loewy, *Never Leave Well Enough Alone*, 131.

p. 66: **"Well, young man, what can you do for this railroad?":** Loewy, *Never Leave Well Enough Alone*, 135.

p. 66: **"Apparently, he kept a pleasant memory of the experience":** Loewy, *Never Leave Well Enough Alone*, 135.

p. 67: **He spent three days "looking over the trash can situation":** Loewy, *Never Leave Well Enough Alone*, 137.

Chapter 5. Big Engines

Sources for Raymond Loewy's work with the Pennsylvania Railroad include *Industrial Design*; *Never Leave Well Enough Alone*; *Raymond Loewy* by Paul Jodard; and *Raymond Loewy* by Glenn Porter. The section on Donald Dohner's work on the GG-1 came from a seminal article in *Classic Trains* magazine. Sources for other designers' work on railroad designs include *Henry Dreyfuss, Industrial Designer* by Russell Flinchum; *The Streamlined Era* by Robert Reed; *My Iron Journey* by Otto Kuhler; and *The Pennsylvania Railroad 1940s, 1950s* by Dan Ball Jr. The author's interviews with a variety of Pennsylvania railroad workers were also illuminating, including talks with E. E. "Mac" McIntire, Michael Leberfinger, Rex Bathurst, and Cecil Snyder.

p. 69: **"On a straight stretch of track":** Loewy, *Industrial Design*, 90.

p. 73: **centrally positioned "steeple cab":** Hampton C. Wayt, "Donald Dohner: 'The Man Who Designed Rivets,'" *Classic Trains*, Summer 2009.

p. 74: **"I was thinking in terms of simplification":** Porter, *Raymond Loewy*, 56.

p. 76: **"One thing led to another":** Porter, *Raymond Loewy*, 41.

p. 76: **"We received no authorization from the Pennsylvania railroad":** Glenn Porter, "Troubled Marriage: Raymond Loewy and the Pennsylvania Railroad," *American Heritage*, Spring 1996.

p. 78: **giving the engine a sleek look of an "inverted bathtub":** Flinchum, *Henry Dreyfuss, Industrial Designer*, 63.

p. 79: **"I discovered how to catch a cold in a few minutes":** Loewy, *Industrial Design*, 79.

p. 83: **"They chose a special crew of men—hand-picked—that they called the 'blue-ribbon gang'":** Interview with retired railroader E. E. "Mac" McIntire, Apr. 14, 1988.

p. 83: **"bright as a gun barrel":** Interview with retired railroader Michael Leberfinger, Apr. 1, 1988.

p. 83: **"They wanted everything polished":** Interview with retired railroader Rex Bathurst, Apr. 7, 1988.

p. 84: **Keller Barry, a machinist at the Altoona shops:** Interviews with various railroaders, Altoona, Apr. 22, 1988.

p. 84: **"It sounded just like a jet airplane":** Interview with Cecil Snyder, Apr. 4, 1988.

p. 86: **"Rugged Moderne":** Jodard, *Raymond Loewy*, 55.

p. 86: **"Beautifully streamlined, with an impressive sharknose":** *My Iron Journey*, 182.

p. 86: **The railroad would go on to build fifty of these large duplex engines:** Ball, *Pennsylvania Railroad*, 181.

p. 89: **"I went back to New York totally in disgust":** Special Raymond Loewy Issue, *ID Magazine* (Nov.–Dec. 1986).

p. 90: **interiors seemed to be designed for "vegetarian bacteriologists":** Bayley and Conran, *Design*, 35.

p. 90: **"My youth was charmed by the glamour of the locomotive":** Loewy, *The Locomotive*, 6.

Chapter 6. Constructing an Image while Building a Business

Sources that proved invaluable in describing how Raymond Loewy started his agency include Loewy's *Industrial Design* and *Never Leave Well Enough Alone*; *Raymond Loewy* by Paul Jodard; and *Raymond Loewy* edited by Angela Schoenberger. Recollections in the Loewy memorial issue of *ID Magazine* were crucial in understanding how his offices worked. Sources for other designers' work in building their businesses include *Henry Dreyfuss, Industrial Designer* by Russell Flinchum; *Design This Day* by Walter Teague; and *Designing for People* by Henry Dreyfuss.

p. 91: **"So I rented a swank office on the fifty-fourth floor":** Loewy, *Never Leave Well Enough Alone*, 122.

p. 91: **"Always live better than your clients":** Isadore Barmash, *Always Live Better Than Your Clients*, 9.

p. 91: **"Regardless of what people may tell us, or experts may write":** Loewy, *Never Leave Well Enough Alone*, 122.

p. 92: **"A fellow of taste, he is not only talented but everybody likes him":** Loewy, *Never Leave Well Enough Alone*, 152.

p. 93: **"During the working day he toured the offices":** Elizabeth Reese, "Design and the American Dream," in Schoenberger, *Raymond Loewy*, 39.

p. 94: **"Nobody can sell industrial design but an industrial designer":** Raymond Loewy, *Industrial Design*, 30.

p. 95: **"We can't spend all day on a single drawing"**: Elizabeth Reese, "Design and the American Dream," in Schoenberger, *Raymond Loewy*, 42.

p. 95: **"Loewy prepared for his presentations like a boxer preparing for a big match"**: Elizabeth Reese, "Design and the American Dream," in Schoenberger, *Raymond Loewy*, 42.

p. 95: **"Loewy combined in one stage-managed personality the flair of Bel Geddes and the practicality of Dreyfuss"**: Bayley and Conran, *Design*, 41.

p. 95: **"The best designs always ended up in the wastebasket"**: Elizabeth Reese, "Design and the American Dream," in Schoenberger, *Raymond Loewy*, 43.

p. 96: **"Loewy was a great manipulator"**: Elizabeth Reese, "Design and the American Dream," in Schoenberger, *Raymond Loewy*, 43–44.

p. 96: **"as an expression of confidence and gratitude to my key men"**: Loewy, *Never Leave Well Enough Alone*, 151.

p. 96: **"I started working for Loewy as an office boy"**: Jay Doblin, Special Raymond Loewy Issue, *ID Magazine* (Nov.–Dec. 1986).

p. 98: **"I never got credit for anything I did at Raymond Loewy, nor did anybody else"**: Jodard, *Raymond Loewy*, 65.

p. 98: **"He made it possible for us to work on marvelous accounts"**: Jay Doblin, Special Raymond Loewy Issue, *ID Magazine* (Nov.–Dec. 1986): 43.

p. 98: **"My initial impression of the man was that he was terribly refined and elegant"**: Betty Reese, Special Raymond Loewy Issue, *ID Magazine* (Nov.–Dec. 1986): 40.

p. 98: **"I was a newspaperman's daughter, so I grew up learning a lot of the tricks of the trade"**: Betty Reese, Special Raymond Loewy Issue, *ID Magazine* (Nov.–Dec. 1986): 40.

p. 99: **Loewy described his idea for the room as "a clinic"**: Jodard, *Raymond Loewy*, 65.

p. 100: **typical children's rooms were "grossly over-decorated"**: Porter, *Raymond Loewy*, 50.

p. 100: **"The designer believes that the child"**: Porter, *Raymond Loewy*, 50.

p. 101: **Pepsodent tube "modeled by Loewy"**: "Up from the Egg," *Time*, Oct. 31, 1949.

p. 101: **"He was, when he wanted to be, irresistibly charming"**: John Ebstein, Special Raymond Loewy Issue, *ID Magazine* (Nov.–Dec. 1986): 39.

p. 102: **"Many designers were happy for the experience"**: John Ebstein, Special Raymond Loewy Issue, *ID Magazine* (Nov.–Dec. 1986): 39.

p. 102: **"Our whole economy is based on planned obsolescence"**: Karl Prentiss, "Brook Stevens Interview," *True Magazine* (Apr. 1958).

p. 102: **"Loewy made sure to orchestrate the clients' first view of his designs"**: Betty Reese, Special Raymond Loewy Issue, *ID Magazine* (Nov.–Dec. 1986).

p. 103: **"When I first met Raymond, I found him much too flashy"**: Evert Endt, "A Frenchman in New York," in Schoenberger, *Raymond Loewy*, 30.

p. 104: **"No sounding of horns, no brake screeches"**: Loewy, *Never Leave Well Enough Alone*, 156.

p. 104: **"We never lose contact with reality, and we do not underestimate our social responsibilities"**: Loewy, *Never Leave Well Enough Alone*, 157.

p. 106: **"Franz [Wagner] has a flair for organization"**: Loewy, *Never Leave Well Enough Alone*, 152.

p. 107: **"Some people who know him well suspect that he is not unaware of his brilliance"**: Loewy, *Never Leave Well Enough Alone*, 152.

p. 107: **"Another difference between the two men"**: Betty Reese, Special Raymond Loewy Issue, *ID Magazine* (Nov.–Dec. 1986): 42.

p. 108: **"The store itself becomes American Suburbia's village green"**: Jodard, *Raymond Loewy*, 119.

p. 108: **department stores were "probably the last of the manually operated large industries"**: Loewy, *Never Leave Well Enough Alone*, 200.

p. 112: **Membership was confined to practitioners or professors of industrial design:** Loewy, *Never Leave Well Enough Alone*, 183.

p. 112: **"Our small group became appalled and a bit frightened by this stampede led by fast-buck artists"**: Loewy, *Industrial Design*, 36.

p. 112: **"During the late forties and early fifties"**: Jay Doblin, Special Raymond Loewy Issue, *ID Magazine* (Nov.–Dec. 1986): 44.

Chapter 7. Engines of Industry

The books and other sources used to describe the work of Raymond Loewy and other designers at the 1939 World's Fair include Loewy's *Industrial Design* and *Never Leave Well Enough Alone*; *Raymond Loewy* by Paul Jodard; and *Raymond Loewy* edited by Angela Schoenberger. Recollections in the Loewy memorial issue of *ID Magazine* were crucial in understanding how his offices worked. Sources for other designers' work in building their businesses include *Henry Dreyfuss, Industrial Designer* by Russell Flinchum; *Design This Day* by Harold Van Doren; *Designing for People* by Henry Dreyfuss; *Objects of Desire* by Adrian Forty; and *Raymond Loewy* by Glenn Porter. Passages on the intersection of advertising and industrial design were gleaned from *The Mirror Makers* by Stephen Fox. Descriptions of Loewy's work on International Harvester came from *150 Years of International Harvester* by C. H. Wendel, and his work on Greyhound buses stems from *Hounds of the Road* by Carlton Jackson and *One Hundred Great Product Designs* by Jay Doblin. Jeffrey Meikle's *Twentieth Century Limited* and Donald Bush's *The Streamlined Era* also were invaluable resources.

p. 115: **The machines created extra income:** Wendel, *150 Years of International Harvester*, 89.

p. 117: **"I left Chicago for Fort Wayne [Indiana] on the train"**: Loewy, *Industrial Design*, 126.

p. 118: **"The design contradicts [the idea] that trademarks demand thorough, lengthy, expensive research"**: Loewy, *Industrial Design*, 126.

p. 119: **The company was founded in 1914 as a local bus line in Hibbing, Minnesota:** Jackson, *Hounds of the Road*, 10.

p. 120: **drivers did not particularly relish the idea of driving a dachshund:** Jackson, *Hounds of the Road*, 19.

p. 120: **an animal Loewy referred to as a "fat mongrel":** Jodard, *Raymond Loewy*, 98.

p. 122: **"ferry transport of tomorrow, today":** Bush, *Streamlined Decade*, 49.

p. 122: **"When illuminated at night, the *Princess Anne* looked like a giant liner":** Loewy, *Industrial Design*, 95.

p. 123: **"We are not going to reproduce any classic styles; these ships will be entirely modern":** Porter, *Raymond Loewy*, 46.

p. 124: **Formica tabletops set in tubular metal frames:** Meikle, *Twentieth Century Limited*, 113.

Chapter 8. Studebaker Beginnings

A variety of books and sources were used in reconstructing Raymond Loewy's long career with Studebaker. Loewy's *Industrial Design* and *Never Leave Well Enough Alone*; *Raymond Loewy* by Paul Jodard; and *Raymond Loewy* edited by Angela Schoenberger were all immensely useful as references and sources. The automotive sources used for this chapter included *Billy, Alfred, and General Motors* by William Pelfrey; *The Art of American Car Design* by C. Edson Armi; *More Than They Promised* by Thomas Bonsall; *Studebaker* by Donald T. Critchlow; and *Studebaker* by Michael Beatty, Patrick Furlong, and Loren Pennington. The sources used for automotive styling history were *Auto-Opium* by David Gartman; *A Century of Automotive Style* by Michael Lamm and Dave Holls; and *The Story of Paul G. Hoffman* by Crane Haussaman. The reminiscences of various Studebaker designers came from the Autolife Archive at University of Michigan, as well such books as *Studebaker* by Richard M. Langworth and *Virgil Exner* by Peter Grist.

p. 125: **"The automobile was an invention, and it looked like one":** Raymond Loewy, "Jukebox on Wheels," *Atlantic* (Apr. 1955): https://www.theatlantic.com/magazine/archive/1955/04/jukebox-on-wheels/303944/.

p. 125: **"Using his own funds, he designed":** See the Avanti website: www.theavanti.com.

p. 128: **At the turn of the twentieth century, there were 270 automobile companies operating nationally:** Critchlow, *Studebaker*, 49.

p. 129: **"Studebaker didn't cut much of a figure in the auto business":** Critchlow, *Studebaker*, 55.

p. 129: **"Billy was smart and had the marks of a gentleman":** Bonsall, *More Than They Promised*, 64.

p. 130: **"our men build their very souls into the Studebaker cars":** Beatty et al., *Studebaker*, 15.

p. 130: **"Sometimes I used to feel":** Pelfrey, *Billy, Alfred, and General Motors*, 12.

p. 131: **"when Mr. Durant visited":** Pelfrey, *Billy, Alfred, and General Motors*, 146.

p. 133: **"styling, with improved function, would not only sell well"**: Raymond Loewy, "Jukebox on Wheels," *Atlantic* (Apr. 1955): https://www.theatlantic.com/magazine/archive/1955/04/jukebox-on-wheels/303944/.

p. 133: **"every time you get in it, it's a relief—you have a little vacation for a while"**: Gartman, *Auto-Opium*, 94.

p. 134: **"He had charisma in spades"**: Armi, *Art of American Car Design*, 34.

p. 134: **"Almost every designer responsible for the shape of tomorrow's car"**: Raymond Loewy, "Car of the Future," *Science and Mechanics* (Aug. 1950).

p. 135: **"When he wanted something it would be very difficult to work with him because he knew what he wanted and he couldn't draw it for you"**: Armi, *Art of American Car Design*, 223.

p. 136: **"Harley was always so image-conscious"**: Armi, *Art of American Car Design*, 20.

p. 136: **"Earl wasn't a designer himself. But he was one of the finest critics of design ever to come along"**: Armi, *Art of American Car Design*, 25.

p. 137: **"He was a very strong, big man"**: Armi, *Art of American Car Design*, 27.

p. 137: **"Our father who art in styling, Harley be thy name"**: Armi, *Art of American Car Design*, 32.

p. 137: **"They led the life of jetsetters and designed cars with the certainty and abandon of demigods"**: Armi, *Art of American Car Design*, 47.

p. 137: **"Harley Earl designed a car so that when you walked around it, you'd be entertained the whole trip"**: Lamm and Holls, *Century of Automotive Style*, 97.

p. 137: **"Big companies jumped on the 'style' bandwagon"**: Raymond Loewy, "Jukebox on Wheels," *The Atlantic* (Apr. 1955): https://www.theatlantic.com/magazine/archive/1955/04/jukebox-on-wheels/303944/.

p. 139: **The plant, featuring 7.5 million feet of covered floor space on 126 acres:** "Low-Slung Beauty," *Time*, Feb. 2, 1953.

p. 140: **"A man coming into the sales room should always be met with the idea"**: Haussaman, *Story of Paul G. Hoffman*, 8.

p. 141: **"[We were going to] bring out the Rockne, which was our challenge to Ford"**: Critchlow, *Studebaker*, 55.

p. 142: **"we felt that our own designers were paying too much attention to the production engineers"**: "Low-Slung Beauty," *Time*, Feb. 2, 1953.

p. 142: **"Each one of us was given a number for our designs"**: David Crippen, interview with Audrey Moore Hodges, Autolife Archive, University of Michigan.

p. 142: **Loewy "was a very fine person to work with, and he was so chic"**: David Crippen, interview with Audrey Moore Hodges, Autolife Archive, University of Michigan.

p. 143: **"I tried hard to convince management, as early as 1939"**: Langworth, *Studebaker*, 22.

p. 144: **"The astonishingly spare use of brightwork":** Bonsall, *More Than They Promised*, 204.

p. 145: **For years afterward, the word "Studebaker" meant "truck" in Russian slang:** Critchlow, *Studebaker*, 120.

p. 146: **"We designed in three dimensions there":** David Crippen, interview with Robert Andrews, Autolife Archive, University of Michigan.

p. 146: **"He has so many ideas, he didn't know when to turn them off":** Bridges, *Bob Bourke's Designs for Studebaker*, 43.

p. 146: **the board sought to even the postwar playing field by decreeing:** Jodard, *Raymond Loewy*, 81.

p. 148: **"He does not embellish or elaborate, but refines, simplifies and perfects":** Armi, *Art of American Car Design*, 55.

p. 148: **"Those cliff dwellers didn't like cars":** Armi, *Art of American Car Design*, 217.

p. 148: **"Loewy lowered his lance against two All-American fetishes":** Jay Doblin, Special Raymond Loewy Issue, *ID Magazine* (Nov.–Dec. 1986).

p. 148: **"One of my main and generally little known contributions to Studebaker was my frequent presence in Europe":** Langworth, *Studebaker*, 24.

p. 149: **"There is much to be gained working backward from optimal form to mechanics":** Armi, *Art of American Car Design*, 55.

p. 149: **Bourke started his career at Sears, "designing such memorable devices as manure spreaders":** Langworth, *Studebaker*, 32.

p. 149: **"I was a very naïve young man":** Buehrig and Jackson, *Rolling Sculpture*, 17.

p. 151: **"Virg had been there before I was, and Virg was my assistant":** David Crippen, interview with Gordon Buehrig, Autolife Archive, University of Michigan.

p. 151: **Exner, "a man of immense ego," was also having trouble ceding design credit to Loewy:** Critchlow, *Studebaker*, 130.

p. 151: **"Loewy would show up out there from time to time, and Ex would mumble under his breath":** David Crippen, interview with Robert Bourke, Autolife Archive, University of Michigan.

p. 151: **"I'd always admired him—especially his ability to sell advanced designs to recalcitrant executives":** Langworth, *Studebaker*, 15.

p. 151: **"Ex felt a man was either a designer or a promoter":** Langworth, *Studebaker*, 24.

p. 152: **"We had mutual respect for each other":** Buehrig and Jackson, *Rolling Sculpture*, 117.

p. 152: **"You have to know them all—the histories":** Armi, *Art of American Car Design*, 218.

p. 153: **"I agreed to start immediately":** Grist, *Virgil Exner*, 37.

p. 154: **"The trouble was promoted by Roy Cole because even my father had trouble getting Loewy out from New York":** David Crippen, interview with Virgil Exner Jr., Autolife Archive, University of Michigan.

p. 154: **"For several months we were working individually on the package":** Bonsall, *More Than They Promised*, 243.

p. 155: **"I soon realized that Exner's conception of advanced body styling clashed with the ideas"**: Langworth, *Studebaker*, 26.

p. 155: **"I will have nothing more to do with you"**: Grist, *Virgil Exner*, 39.

p. 155: **"In my experience in fifty years, working in more than one hundred corporations"**: Bonsall, *More Than They Promised*, 244.

p. 155: **"You are immediately hired Mr. Exner, by the Studebaker Corporation"**: Grist, *Virgil Exner*, 39.

p. 155: **"It was sort of an underhanded deal on the part of Roy Cole"**: Langworth, *Studebaker*, 27.

p. 155: **Loewy "was very temperamental and these other people were temperamental"**: David Crippen, interview with Audrey Moore Hodges, Autolife Archive, University of Michigan.

p. 156: **"We had to wheel our full-size clay models through the streets of South Bend"**: Bonsall, *More Than They Promised*, 208.

p. 156: **"We would be invited to a showing of one of our designs"**: David Crippen, interview with Audrey Moore Hodges, Autolife Archive, University of Michigan.

p. 156: **"He couldn't draw cars. He was just ridiculous and he knew it"**: Lamm and Holls, *Century of Automotive Style*, 210.

p. 156: The new postwar Studebaker model was the **"great leap forward"**: Lamm and Holls, *Century of Automotive Style*, 210.

p. 156: **"Studebaker made Loewy a household word"**: Lamm and Holls, *Century of Automotive Style*, 205.

p. 157: **"The management's decision to introduce genuine postwar models as quickly as possible"**: Bonsall, *More Than They Promised*, 246.

p. 157: **"I wasn't for chrome either"**: Armi, *Art of American Car Design*, 79.

p. 157: **"The independent, in order to succeed, must be courageous and progressive"**: Armi, *Art of American Car Design*, 79–80.

p. 157: **"Raymond Loewy's first postwar Studebaker was a car that reflected its honesty"**: Dreyfuss, *Designing for People*, 141.

p. 158: **"The bullet-nose Studebaker is almost unique among his production cars"**: Bonsall, *More Than They Promised*, 258.

p. 158: **"The result was a bulbous, rather clumsy, fat automobile"**: Bonsall, *More Than They Promised*, 258.

p. 158: **"Still there is one undeniable forerunner of the bullet-nose concept"**: Bruno Scacco, "The Studebaker Connection," in Schoenberger, *Raymond Loewy*, 129.

p. 159: **"Boy, if they ever come out with a silly thing like that we'll go bankrupt"**: Critchlow, *Studebaker*, 131.

p. 160: **"The 1947 Studebaker bore the clean imprint of Raymond Loewy's designing genius"**: John Cooper Fitch, "10 for the Road," *Esquire* (Dec. 1960).

p. 161: **Caleal's Ford design was referred to as "the car that saved an empire"**: Gantz, *Founders of American Industrial Design*, 140.

p. 161: **"We smoothed those lines out and began the movement toward integra-tion of the fenders and the body":** David Crippen interview with George Walker, Autolife Archive, University of Michigan.

p. 161: **"He did so little. He did clay modeling for us, that's what he did":** David Crippen interview with George Walker, Autolife Archive, University of Michigan.

p. 161: **"Dick had plenty of help":** David Crippen interview with Frank Bianchi, Autolife Archive, University of Michigan.

p. 161: **"[Bob] Koto worked on the sides and the roof, and he worked on the windshield, the window details":** David Crippen interview with Robert Bourke, Autolife Archive, University of Michigan.

p. 162: **"We wanted in the Forward Look cars an appearance of fleetness":** Grist, *Virgil Exner*, 81.

p. 163: **"My God Mildred! They want to cut [the fins] because they can save 36 cents!":** Jason Stein, "Exner Gained Fame as the "Fin Man," *Hagerstown Herald-Mail*, July 9, 2006.

p. 163: **"Thanks to Paul Hoffman, I was given the opportunity to design cars liberated from most of Detroit's atavistic style":** Loewy, *Industrial Design*, 148.

Chapter 9. The Starliner Coupe

Most of the sources from chapter 8 were also invaluable in tracing Raymond Loewy's Studebaker work. Of particular utility was the February 2, 1953, issue of *Time* detailing the debut of the Starliner; *The American Design Adventure, 1940–1975* by Arthur J. Pulos; and *Bob Bourke's Designs for Studebaker* by John Bridges. The section on Brooks Stevens is indebted to *Industrial Strength Design* by Glenn Adamson.

Loewy's *Industrial Design* and *Never Leave Well Enough Alone*; *Raymond Loewy* by Paul Jodard; and *Raymond Loewy* edited by Angela Schoenberger were all immensely useful as references and sources. The automotive sources used for this chapter included *The Art of American Car Design* by C. Edson Armi; *More Than They Promised* by Thomas Bonsall; *Studebaker* by Donald T. Critchlow; and *Studebaker* by Michael Beatty, Patrick Furlong, and Loren Pennington. The sources used for automotive styling history were *A Century of Automotive Style* by Michael Lamm and Dave Holls; and *The Story of Paul G. Hoffman* by Crane Haussaman. The reminiscences of various Studebaker designers came from the Autolife Archive at University of Michigan, as well as from *Studebaker* by Richard M. Langworth.

p. 164: **"For six months the car was driven, in well-shrouded secrecy":** "Low-Slung Beauty," *Time*, Feb. 2, 1953.

p. 166: **"I always wanted the chance to show what could be done if we didn't have too many restrictions":** Bridges, *Bob Bourke's Designs for Studebaker*, 106.

p. 166: **"We knew it would [sell] if it would be fresh and gay and young-looking":** "Low-Slung Beauty," *Time*, Feb. 2, 1953.

p. 167: **"The 1953 Studebakers could not have been produced without a total consensus"**: Bruno Sacco, "The Studebaker Connection," in Schoenberger, *Raymond Loewy*, 129.

p. 169: **A few weeks later, as a "show of force"**: Bridges, *Bob Bourke's Designs for Studebaker*, 106.

p. 170: **"The reason it was exciting is that Mr. Loewy had the only international design department in the United States"**: Robert Andrews, interview by David Crippen, Autolife Archive, University of Michigan.

p. 170: **"I had felt for the last forty years that the American automobile was too bulky and heavy"**: Loewy, *Industrial Design*, 148.

p. 170: **"Detroit simply had no use for my ideas"**: Loewy, *Industrial Design*, 148.

p. 172: **"We practical designers deplore just as much as the engineers the excessive bulk and weight"**: Loewy, *Never Leave Well Enough Alone*, 307.

p. 173: **"Unfortunately Studebaker was not big enough a car company to set automotive styles"**: Hine, *Populuxe*, 95.

p. 173: **"His talent was selling the design to top management, a talent most designers lack"**: Bridges, *Bob Bourke's Designs*, 41.

p. 173: **"He would sketch to some degree with a soft pencil in a heavy-line style"**: Bridges, *Bob Bourke's Designs*, 41.

p. 175: **He called Detroit designs "cars filled with spinach and schmaltz"**: Raymond Loewy, "Jukebox on Wheels," *The Atlantic* (April 1955): 36.

p. 176: **"is it responsible to camouflage one of America's most remarkable machines"**: Hine, *Populuxe*, 98.

p. 176: **"Detroit's reliance on the stylist is based on the perverse notion"**: Henry Dreyfuss, "The Car Detroit Should Be Building," *Consumer Reports* (July 1958).

p. 176: **The Packard takeover was a universal cause for resentment**: Beatty et al., *Studebaker*, 34.

p. 177: **"Nance was one of the worst automobile executives I've ever come in contact with"**: L. David Ash, interview by David Crippen, Autolife Archive, University of Michigan.

p. 177: **"Our billings were about [$1 million] a year"**: Langworth, *Studebaker*, 81.

p. 178: **"It took Nance something like a year and a half to go through $4 million"**: Bridges, *Bob Bourke's Designs*, 41–42.

p. 178: **"Nearly seven years ago, when the company could well afford"**: Raymond Loewy to Paul Hoffman or Harold Churchill, Raymond Loewy Archive, Hagley Museum and Library.

p. 179: **"I am told that cab drivers have the highest rate of duodenal ulcers"**: Raymond Loewy, "Jukebox on Wheels," *The Atlantic* (April 1955).

p. 182: **creating "a spiffy-looking switch box"**: Adamson, *Industrial Strength Design*, 2.

p. 182: **"I had to fight my way in to talk to anybody in the '30s"**: John Holusha, "Brooks Stevens, 83, Giant in Industrial Design," *New York Times*, Jan. 7, 1995.

p. 182: **"I decided at that point that an individual crusade":** "Planned Obsolescence," *Finish Magazine* (Sept. 1956): 12.

p. 183: **Some wags called the home "the only Greyhound Bus terminal in Fox Point":** Adamson, *Industrial Strength Design*, 62.

p. 183: **"We were out to sell boats to coal miners and lathe operators":** Adamson, *Industrial Strength Design*, 28.

Chapter 10. Avanti

The story of Raymond Loewy and the Avanti was gathered from the following sources: Loewy's *Industrial Design*; *Raymond Loewy* by Paul Jodard; and *Raymond Loewy* edited by Angela Schoenberger, which provided invaluable background on this unique sports car. The automotive sources used for this chapter included *The Art of American Car Design* by C. Edson Armi; *More Than They Promised* by Thomas Bonsall; *Studebaker* by Donald T. Critchlow; and *Studebaker* by Michael Beatty, Patrick Furlong, and Loren Pennington. *A Century of Automotive Style* by Michael Lamm and Dave Holls was a source for automotive styling history. The website www.theavanti .com provided excellent resources for the marketing of the Avanti and the design details of the production car. The reminiscences of various Studebaker designers came from the Autolife Archive at University of Michigan, as well as from *Studebaker* by Richard M. Langworth. Critiques of the Avanti were found in *In Good Shape* by Stephen Bayley, and the car's complicated history was best told in *Avanti* by John Hull. The reminiscences of Raymond Loewy's daughter, the late Laurence Loewy, come from the website she oversaw until her death, www.raymondloewy.com.

p. 186: **"He handed me a bunch of clippings about cars which he's been carrying around":** Bonsall, *More Than They Promised*, 359.

p. 187: **He was looking for "a single dramatic new model":** Bonsall, *More Than They Promised*, 358.

p. 187: **"Let me do it the way I want to and give me complete freedom of action":** Langworth, *Studebaker*, 130.

p. 189: **"it's El Morocco without the rhumba band":** Dorothy Kilgallen, "The Voice of Broadway," *New York Journal American*, July 24, 1948.

p. 190: **There is a photograph of Loewy explaining the airfoil to General Charles de Gaulle at a Paris auto show:** Jodard, *Raymond Loewy*, 137.

p. 190: **"The bulbous rear was even more clumsily handled":** Len Frank, "1963 Studebaker Avanti R2: The Car That Even Studebaker Could Not Kill," *Motor Trend* (Feb. 1982).

p. 191: **"I was often frustrated working for Detroit, and sometimes by simply observing its designs":** Loewy, *Industrial Design*, 136.

p. 191: **"Raymond Loewy, Studebaker's designer and chief stylist, proved once again in 1953 that he's the guy the rest of the country's designers wish they were":** Loewy, *Industrial Design*, 138.

p. 192: **Loewy also included a rough sketch of a car with a "Coca-Cola curve":** Bayley, *In Good Shape*, 20.

p. 194: **Bob Andrews described it as being "like a cloak and dagger movie":** Langworth, *Studebaker*, 130.

p. 194: **"Egbert did not try to become an automobile designer overnight, which happens so much":** Robert Andrews interview by David Crippen, Autolife Archive, University of Michigan.

p. 198: **Kellogg's personal letters from 1965 and 1966 put to rest:** Tom Kellogg to Raymond Loewy, 1965 and 1966, Raymond Loewy Archive, Hagley Museum and Library.

p. 199: **"It made the car and driver integral":** Langworth, *Studebaker*, 135.

p. 199: **The brochure for the Avanti, designed by the d'Arcy Advertising Agency:** Sales brochure, Studebaker Motor Company, 1963, http://www.oldcarbrochures.com/new/120311/1963%20Avanti%20Brochure/dirindex.html.

p. 199: **"Its clean crisp lines represent a fresh departure from today's trend toward over-ornamentation":** Sales brochure, Studebaker Motor Company, 1963, http://www.oldcarbrochures.com/new/120311/1963%20Avanti%20Brochure/dirindex.html.

p. 200: **"This will surely interest the ladies":** Sales brochure, Studebaker Motor Company, 1963, http://www.oldcarbrochures.com/new/120311/1963%20Avanti%20Brochure/dirindex.html.

p. 200: **"to reach those indispensable items which clutter up a car's interior":** Sales brochure, Studebaker Motor Company, 1963, http://www.oldcarbrochures.com/new/120311/1963%20Avanti%20Brochure/dirindex.html.

p. 201: **Calling the car a "hot, saleable line":** Hull, *Avanti*, 21.

p. 202: **"There is some concern among Studebaker's conservative engineers":** Joseph Ingraham, "Studebaker Gives Daring Car Plan," *New York Times*, Apr. 10, 1962.

p. 203: **"It was catastrophic":** Langworth, *Studebaker*, 141.

p. 204: **"a car for the driving sport as opposed to the sporting driver":** Langworth, *Studebaker*, 140.

p. 204: **"No one can expose a body to the public":** *Detroit News*, Jan. 15, 1964.

p. 205: **"By the time the Avanti came out, there were very few dealerships":** Robert Andrews interview by David Crippen, Autolife Archive, University of Michigan.

p. 206: **"He was never a man to take small steps":** Bruno Sacco, "The Studebaker Connection," in Schoenberger, *Raymond Loewy*, 133.

p. 206: **"I still keep two beige Avantis, one in Paris, one in Palm Springs":** Loewy, *Industrial Design*, 182.

p. 206: **"The two great experiences":** Raymond Loewy, personal reminiscence, 1970, Raymond Loewy Archive, Hagley Museum and Library.

p. 206: **"My decades with the company were exhilarating and unforgettable":** Loewy, *Industrial Design*, 137.

p. 207: **Studebaker dealers were considered "a complacent and non-competitive group":** Joseph Ingraham, "Studebaker Auto Making Ended Except in Canada," *New York Times*, Dec. 10, 1963.

p. 207: **The sudden closing gave the company the opportunity to cut the pension plan:** Pollack, *Why the PBGC Termination Insurance Program Should Be Ended*.

p. 208: **"The possibility that the South Bend operation could collapse":** Beatty et al., *Studebaker*, 53.

p. 209: **The company ceased production in 2006:** Hull, *Avanti*, 119.

p. 210: **"Did you ever stop to wonder what they did with the lush profits of the war years?":** Hull, *Avanti*, 175.

p. 211: **"The fall of Studebaker was not inevitable":** Bonsall, *More Than They Promised*, 470.

Chapter 11. Becoming a Businessman

The details of how Raymond Loewy ran his business are described in detail and with strikingly good humor in *Never Leave Well Enough Alone*. His later work and business practices after 1960 are covered in detail in *Industrial Design*. The work done by Henry Dreyfuss and Associates is explained in Dreyfuss's *Designing for People*, and Donald Deskey's work is detailed in *Donald Deskey* by David Hanks and Jennifer Toher. Sources for describing how Loewy's agency functioned include *Raymond Loewy* by Paul Jodard and *Raymond Loewy* edited by Angela Schoenberger. The Loewy office was recollected in the Loewy memorial issue of *ID Magazine*, which also detailed his personality and management style. Sources for other designers' work in building their businesses include *Henry Dreyfuss, Industrial Designer* by Russell Flinchum and *Design This Day* by Harold Van Doren.

p. 213: **"They are dropped, banged, hit, pushed, pulled":** Loewy, *Never Leave Well Enough Alone*, 264.

p. 214: **"By the time refreshments have taken effect":** Loewy, *Never Leave Well Enough Alone*, 265.

p. 215: **"There seems to be for each individual product":** Loewy, *Never Leave Well Enough Alone*, 278–79.

p. 215: **"For a big corporation, a little style goes a long way":** Loewy, *Never Leave Well Enough Alone*, 278–79.

p. 215: **"I wouldn't buy that stuff. It's too extreme":** Loewy, *Never Leave Well Enough Alone*, 278–79.

p. 216: **"People are very open-minded about new things":** Loewy, *Never Leave Well Enough Alone*, 278–79.

p. 216: **Dreyfuss's list of priorities for design is markedly shorter and nearly opposite in emphasis:** Dreyfuss, *Designing for People*, 178.

p. 217: **"A case of pernicious sales anemia sets in"**: Loewy, *Never Leave Well Enough Alone*, 281.

p. 218: **"Though it is difficult to place the drawings in a set order"**: Jodard, *Raymond Loewy*, 16.

p. 219: **"The realm of Freudian fetishes"**: John Kobler, "The Great Packager," *Life*, May 2, 1949.

p. 219: **"broods a great deal about the callipygian Coke bottle"**: John Kobler, "The Great Packager," *Life*, May 2, 1949.

p. 220: **"He was also a master manipulator of his staff"**: Betty Reese, Special Raymond Loewy Issue, *ID Magazine* (Nov.–Dec. 1986): 41.

p. 221: **"Loewy adored draftsmen. He would make a little sketch that was really quite primitive"**: Betty Reese, Special Raymond Loewy Issue, *ID Magazine* (Nov.–Dec. 1986): 42.

p. 221: **"His people came to think of themselves as the design elite in their various specialties"**: Elizabeth Reese, "Design and the American Dream," in Schoenberger, *Raymond Loewy*, 39.

p. 222: **"By giving the men a chance to escape their regular surroundings, to work at a different pace, with different associates, their outlooks are refreshed"**: Loewy, *Never Leave Well Enough Alone*, 309.

p. 222: **"Industrial design, as far as I'm concerned is 25 percent inspiration and 75 percent transportation"**: Loewy, *Never Leave Well Enough Alone*, 321.

p. 222: **"No one had much contact with the man"**: Jay Doblin, Special Raymond Loewy Issue, *ID Magazine* (Nov.–Dec. 1986): 43.

p. 222: **The firm was a "benign dictatorship"**: Elizabeth Reese, "Design and the American Dream," in Schoenberger, *Raymond Loewy*, 39.

p. 223: **"Loewy's insistence on the importance and relevance of industrial design"**: Jodard, *Raymond Loewy*, 172–73.

p. 225: **"over walls made of egg-crate fiber"**: "Up from the Egg," *Time*, Oct. 31, 1949.

p. 226: **an industrial designer "was no living embodiment of Leonardo Da Vinci"**: Hanks and Toher, *Donald Deskey*, 124.

p. 227: **"I felt very happy, and I walked a great deal through the desert at the foot of Mount Jacinto"**: Loewy, *Never Leave Well Enough Alone*, 350.

p. 227: **a house "for a solitary man" . . . "He didn't even have a phone"**: Nicolai Ouroussoff, "House Proud: Retrieving the Future Under a Desert Sky," *New York Times*, Dec. 28, 1995.

p. 228: **"It was an interesting collaboration with Loewy because he had wonderfully inventive ideas"**: *New York Times Magazine*.

p. 229: **"What I liked most was the subtlety of the coloring"**: Loewy, *Never Leave Well Enough Alone*, 350.

p. 229: **"I called on a local architect of whom I had heard"**: Loewy, *Never Leave Well Enough Alone*, 351.

p. 229: **"Jack Benny likes to spend a quiet hour with Viola and me"**: Loewy, *Never Leave Well Enough Alone*, 353.

p. 229: **"When the house is kept in darkness":** Loewy, *Never Leave Well Enough Alone*, 352.

p. 230: **"We've always lived life to the hilt":** Susan Heller Anderson, "The Pioneer of Streamlining: Design," *New York Times*, Nov. 4, 1979.

p. 230: **"We are surrounded by our friends":** Loewy, *Never Leave Well Enough Alone*, 354.

p. 231: **"I erected a house I actually designed in New York":** Loewy, *Industrial Design*, 44.

p. 231: **"America gives me the opportunity to be creative and imaginative":** Loewy, *Never Leave Well Enough Alone*, 357.

Chapter 12. The Sales Curve Wanes

Industrial Design and *Never Leave Well Enough Alone* provided insight into Raymond Loewy's business history. Paul Junod's *Raymond Loewy* and Arthur Pulos's *American Design Adventure* detailed how Loewy and his contemporaries built the industrial design profession. Two other books, *Raymond Loewy* edited by Angela Schoenberger and *Donald Deskey* by David A. Hanks and Jennifer Toher, were excellent sources. Jay Doblin's *100 Great Product Designs* and *Raymond Loewy and Streamlined Design* by Philippe Tretiak filled in critical parts of this aspect of Loewy's career.

p. 233: **"In creating the container, I like to think I did something for American men as well!":** Loewy, *Industrial Design*, 52.

p. 234: **"There were giants around the table":** Loewy, *Never Leave Well Enough Alone*, 181.

p. 234: **"I remember vividly the feeling of pride":** Pulos, *American Design Adventure*, 200.

p. 235: **A designer was defined as one "who has successfully designed":** Jodard, *Raymond Loewy*, 106.

p. 235: **"Captives! Following the orders of marketing managers":** Jodard, *Raymond Loewy*, 108.

p. 236: **"If designers get reabsorbed, digested, ingested, mutated or reoriented":** Pulos, *American Design Adventure*, 208.

p. 237: **"I would not call him a rounded man":** Fox, *Mirror Makers*, 115.

p. 238: **"What about that package? Do you really believe you could improve on it?":** Loewy, *Never Leave Well Enough Alone*, 146.

p. 238: **"The public's reaction to entertainment and advertising is no different":** Fox, *Mirror Makers*, 149–50.

p. 238: **Hill accepted the "bet," offering Loewy a $20,000 retainer:** Loewy, *Never Leave Well Enough Alone*, 146.

p. 238: **"Oh, I don't know, some nice spring morning":** Loewy, *Never Leave Well Enough Alone*, 147.

p. 238: **"Putting the logo on both sides of the box was Loewy's genius":** Tretiak, *Raymond Loewy and Streamlined Design*, 10.

p. 239: **"Hygiene, cleanliness and comfort are uniquely American product needs":** Forty, *Objects of Desire*, 243.

p. 244: **the "Sheer Look":** Doblin, *One Hundred Great Product Designs*, 107.

p. 244: **"Throughout history, works of art and purely functional objects retain their cultural and material value":** George Higby, "Rosenthal China's Modernist Masters," *Antiques Quarterly* (Winter 2008): http://www .myantiquemall.com/rosenthaldinnerwaresetvintageantiqueporcelain /rosenthaldinnerwaresetvintageantiqueporcelain.html.

p. 247: **designers moved "From Celebrity to Anonymity" during the 1960s:** Jeffrey L. Meikle, "From Celebrity to Anonymity," in Schoenberger, *Raymond Loewy*, 51.

p. 249: **Loewy's "restless versatility":** Jodard, *Raymond Loewy*, 156.

Chapter 13. The Long Road Down

The sources referencing Loewy's final working years and retirement were invaluable in completing an often murky picture. *In Good Shape* by Stephen Bayley and *Raymond Loewy and Streamlined Design* by Philippe Tretiak were effective in understanding Raymond Loewy's persona. Loewy's *Industrial Design*; *Raymond Loewy* edited by Angela Schoenberger; and *Twentieth-Century Style and Design* by Stephen Bayley and Terence Conran were exceptional sources as well. The discussion of Loewy's work on *Air Force One* could not have been possible without the help of *Air Force One* by Kenneth T. Walsh and Phil Patton's "Air Force One: The Graphic History," an article in the *AIGA Journal of Design*. The section on Loewy's NASA commissions was helped by *Living and Working in Space* by W. David Compton and Charles D. Benson.

p. 251: **"Loewy was an import":** Stephen Bayley, "Public Relations or Industrial Design," in Schoenberger, *Raymond Loewy*, 48.

p. 251: **Loewy's memoir is self-consciously perfect:** Tretiak, *Raymond Loewy and Streamlined Design*, 15.

p. 251: **like a "glamorized movie star—too glamorized":** Tretiak, *Raymond Loewy and Streamlined Design*, 5.

p. 252: **"He was the Leonardo Da Vinci of the New World—both engineer and visionary":** Tretiak, *Raymond Loewy and Streamlined Design*, 12.

p. 252: **They "also opened the floodgates to all sorts of unethical merchants":** Loewy, *Industrial Design*, 36.

p. 255: **"I received a phone call from [McHugh] telling me that the president wished to meet with me":** Loewy, *Industrial Design*, 22.

p. 255: **"In every case I had replaced red . . . with a luminous ultramarine blue":** Loewy, *Industrial Design*, 22.

p. 256: **Loewy is the one on the floor; the president sat in his rocking chair observing:** Phil Patton, "Air Force One: The Graphic History," *AIGA Journal of Design*, Feb. 26, 2009.

p. 256: **White House appointment diaries do not have records of a Loewy visit:** Phil Patton, "Air Force One: The Graphic History," *AIGA Journal of Design*, Feb. 26, 2009.

p. 256: **The typeface for the fuselage lettering was supposedly inspired:** Phil Patton, "Air Force One: The Graphic History," *AIGA Journal of Design*, Feb. 26, 2009.

p. 258: **"TWA pilots were said to boast that their jets seemed to be going 600 miles per hour even when parked":** "Designed to Travel: Curating Relics of T.W.A. as It Prepares for Departure," *New York Times*, June 7, 2001.

p. 258: **"I prepared in my personal design studio":** Raymond Loewy to Francis X. Clair, July 18, 1977, Raymond Loewy Archive, HML.

p. 260: **"the applications and concepts of industrial design":** Loewy, *Industrial Design*, 188.

p. 260: **At 106 feet long and weighing 100 tons, the "baby space station" was designed:** Skylab Special Section, *New York Times*, May 13, 1973.

p. 263: **"Nobody could have lived in that thing for more than two months":** Compton and Benson, *Living and Working in Space*, 133.

p. 264: **"My first opportunity to express a deep interest in the future of spatial exploration":** Reyer Kras, "Loewy Reached for the Stars," in Schoenberger, *Raymond Loewy*, 188.

p. 265: **The Loewy group was assigned to deliver "comments and recommendations":** Compton and Benson, *Living and Working in Space*, 133.

p. 265: **"You and your organization played a crucial role . . ."** Loewy, *Industrial Design*, 205.

p. 266: **"For 35 years I have been privileged to work in close collaboration with top management":** Reyer Kras, "Loewy Reached for the Stars," in Schoenberger, *Raymond Loewy*, 190.

p. 266: **"We can imagine that crew members, feeling encaged in the OWS":** Reyer Kras, "Loewy Reached for the Stars," in Schoenberger, *Raymond Loewy*, 190.

p. 266: **"The porthole which was added at our mutual insistence":** Loewy, *Industrial Design*, 40.

p. 267: **"It is fair to say that we have received sound, professional support from the Loewy-Snaith Company":** Yuri B. Soloviev, "Raymond Loewy in the U.S.S.R.," in Schoenberger, *Raymond Loewy*, 196.

p. 268: **Loewy was able to bring in a commission from the experience:** Memo to Bill Snaith, March 19, 1973, Raymond Loewy Archive, Hagley Museum and Library.

p. 268: **"I was amusing my guests with various anecdotes":** Yuri B. Soloviev, "Raymond Loewy in the U.S.S.R.," in Schoenberger, *Raymond Loewy*, 196.

p. 268: **"The citizens of Lower Slobovia may not give a hoot for freedom of speech":** Frances Stoner Saunders, *The Guardian*, Sept. 6, 2008.

p. 269: **"Not only was Loewy a talented designer, but a brilliant propagandist of design":** Yuri B. Soloviev, "Raymond Loewy in the U.S.S.R.," in Schoenberger, *Raymond Loewy*, 197.

p. 269: **"the first USSR experiment in international cooperation in this field":** "Still at the Drawing Board, Designer Raymond Loewy Shapes Up Russia's Exports," *People*, Mar. 10, 1975.

p. 270: **"Dave: From now on you are taking the responsibilities of an entire body division":** Letters to David Butler, Raymond Loewy Archive, Hagley Museum and Library.

p. 270: **"Of course we have very little of that left":** "Still at the Drawing Board, Designer Raymond Loewy Shapes Up Russia's Exports," *People*, Mar. 10, 1975.

Chapter 14. Legacy

p. 273: **designers "must live internationally to respect and accept the marketplace":** Loewy, *Industrial Design*, 36.

p. 275: **"It is generally accepted that an industrial designer works best when draped at the edge of a turquoise swimming pool":** Loewy, *Never Leave Well Enough Alone*, 306.

p. 275: **"Loewy, who had transformed American icons, ended up crystallized as if in amber, caught in his own trap, transformed into an over-designed product himself":** Tretiak, *Raymond Loewy and Streamlined Design*, 4.

p. 275: **"narcissistic personality disorder defined as a pervasive pattern of grandiosity, need for admiration and lack of empathy":** Bayley and Conran, *Design*, 204.

p. 277: **"I think one real blow to Loewy's popularity with his former employees was that terrible exhibition of his in 1975":** John Ebstein, Special Raymond Loewy Issue, *ID Magazine* (Nov.–Dec. 1986): 42.

p. 278: **"We are the leaders of a modern Crusade":** Text of speech at Renwick Gallery, Aug. 1, 1975, Raymond Loewy Archive, Hagley Museum and Library.

p. 280: **Unfortunately, the Loewy business name survives only as Loewy:** Levent Ozier, "Merger with Seymourpowell Takes Loewy Back to Its Roots in Design and Brand Innovation," Sept. 18, 2007, Dexigner.com.

p. 281: **"I believe in natural talent; if you don't have it, do something else":** Loewy, *Industrial Design*, 15.

p. 285: **"There are not many men whose dying genuinely marks the end of an era":** Introduction, Special Raymond Loewy Issue, *ID Magazine* (Nov.–Dec. 1986): 24.

Bibliography

Adamson, Glenn. *Industrial Strength Design: How Brooks Stevens Shaped Your World*. Cambridge, MA: MIT Press, 2003.

American Association of Museums. *A Selected Collection of Objects from the International Exposition of Modern Decorative and Industrial Art*. Paris: American Association of Museums, 1925.

Armi, C. Edson. *The Art of American Car Design: The Profession and Personalities*. State College: Pennsylvania State University Press, 1989.

———. *American Car Design Now*. New York: Rizzoli, 2003.

Ball, Don, Jr. *The Pennsylvania Railroad 1940s, 1950s*. Chester, VT: Elm Tree Press, 1986.

Barmash, Isadore. *Always Live Better Than Your Clients: The Fabulous Life and Times of Benjamin Sonnenberg, America's Greatest Publicist*. New York: Dodd Mead, 1983.

Bayley, Stephen. *In Good Shape: Style in Industrial Products, 1900–1960*. New York: Van Nostrand Reinhold, 1979.

———. *Harley Earl and the Dream Machine*, New York: Knopf, 1983.

———. *Woman as Design: Before, Behind, Between, Above, Below*. London: Conran Octopus, 2009.

Bayley, Stephen, and Terence Conran. *Design: Intelligence Made Visible*. Richmond Hill, ON: Firefly Books, 2007.

Bayley, Stephen, Philippe Garner, and Deyan Sudjic. *Twentieth Century Style and Design*. New York: Van Nostrand Reinhold, 1986.

Beatty, Michael, Patrick Furlong, and Loren Pennington. *Studebaker: Less Than They Promised*. South Bend, IN: And Books, 1984.

Bel Geddes, Norman. *Horizons*. New York: Little, Brown, 1932.

Bonsall, Thomas E. *More Than They Promised: The Studebaker Story*. Redwood City, CA: Stanford University Press, 2000.

Bridges, John. *Bob Bourke's Designs for Studebaker*. Nashville, TN: J. B. Enterprises, 1984.

Buehrig, Gordon, and William S. Jackson. *Rolling Sculpture: A Designer and His Work*. Newfoundland, NJ: Haessner, 1975.

Bush, Donald. *The Streamlined Decade*. New York: G. Braziller, 1975.

Compton, W. David, and Charles D. Benson. *Living and Working in Space: A History of Skylab*. Mineola, NY: NASA/Dover, 1983.

Critchlow, Donald T. *Studebaker: The Life and Death of an American Corporation*. Bloomington: Indiana University Press, 1996.

Cross, Gary S. *An All-Consuming Century*. New York: Columbia University Press, 2002.

Day, C. R. *Education for the Industrial World: The Ecole d'Arts et Metiers and the Rise of French Industrial Engineering*. Cambridge, MA: MIT Press, 1987.

Doblin, Jay. *One Hundred Great Product Designs*. New York: Van Nostrand Reinhold, 1970.

Dreyfuss, Henry. *Designing for People*. New York: Simon and Schuster, 1955.

Emmet, Boris, and John E. Jueck. *Catalogs and Counters: A History of Sears Roebuck and Company*. Chicago: University of Chicago Press, 1950.

Farr, James, ed. *The Industrial Revolution in Europe, 1750–1914*. Detroit: Thomson/Gale, 2002.

Ferebee, Ann, with Jeff Byles. *A History of Design from the Victorian Era to the Present*. New York: W. W. Norton, 1994.

Flinchum, Russell. *Henry Dreyfuss, Industrial Designer: The Man in the Brown Suit*. New York: Cooper-Hewitt, National Design Museum, Smithsonian Institution and Rizzoli, 1997.

Forty, Adrian. *Objects of Desire*. London: Thames & Hudson, 1986.

Fox, Stephen. *The Mirror Makers: A History of American Advertising and Its Creators*. Champaign: University of Illinois Press, 1997.

Gantz, Carroll. *Founders of American Industrial Design*. Jefferson, NC: McFarland, 2014.

Gartman, David. *Auto-Opium: A Social History of American Automobile Design*. London: Routledge, 1994.

———. *Culture, Class and Critical Theory*. London: Routledge, 2016.

Grist, Peter. *Virgil Exner: Visioneer*. Poundbury, UK: Velece, 2007.

Hanks, David, and Jennifer Toher. *Donald Deskey: Decorative Designs and Interiors*. New York: E. P. Dutton, 1987.

Haussaman, Crane. *The Story of Paul G. Hoffman: A Man Who Shaped His Age*. Santa Barbara, CA: Center for the Study of Democratic Institutions, 1966.

Hine, Thomas. *The Total Package*. New York: Little, Brown, 1995.

———. *Populuxe*. New York: Overlook Books, 2007.

Horne, Alistair. *Seven Ages of Paris*. New York: Knopf, 2002.

Hull, John. *Avanti: The Complete Story*. Hudson, WI: Enthusiast Books, 2008.

Industrial Designers Society of America. *Innovation: Award Winning Industrial Design*. New York: St. Martin's Press, 1997.

Innes, Christopher. *Designing Modern America: From Broadway to Main Street*. New Haven, CT: Yale University Press, 2005.

Jackson, Carlton. *Hounds of the Road: A History of the Greyhound Bus Company*. Bowling Green, OH: Bowling Green University Popular Press, 1984.

Jodard, Paul. *Raymond Loewy*. New York: Taplinger, 1991.

Kuhler, Otto. *My Iron Journey: An Autobiography of a Life with Steam and Steel*. Denver, CO: Intermountain Chapter, National Railway Historical Society, 1967.

Lamm, Michael, and Dave Holls. *A Century of Automotive Style: 100 Years of American Car Design*. Stockton, CA: Lamm-Morada, 1996.

Langworth, Richard M. *Studebaker: 1946–1966 Postwar Years*. Minneapolis: Motorbooks International, 1993.

Leach, William. *Land of Desire: Merchants, Power, and the Rise of a New American Culture*. New York: Pantheon Books, 1994.

Lears, Jackson. *Fables of Abundance: A Cultural History of Advertising in America*. New York: Basic Books, 1995.

Loewy, Raymond. *The Locomotive*. New York: Universe Publishing / Rizzoli, 1937.

———. *Industrial Design*. New York: Overlook Press, 1988. First published 1979.

———. *Never Leave Well Enough Alone*. Baltimore: Johns Hopkins University Press, 2002.

Marchand, Roland. *Advertising the American Dream: Making Way for Modernity 1920–1940*. Berkeley: University of California Press, 1986.

———. *Creating the Corporate Soul: The Rise of Public Relations and Corporate Imagery in Culture, Class and Critical Theory*. Berkeley: University of California Press, 2013.

Marsh, Barbara. *A Corporate Tragedy: The Agony of the International Harvester Company*. New York: Doubleday, 1985.

McMillan, James. *Modern France: 1880–2002*. Oxford: Oxford University Press, 2003.

Meikle, Jeffrey. *Twentieth Century Limited: Industrial Design in America*. Philadelphia: Temple University Press, 1979.

Mierau, Christine. *Accept No Substitutes: The History of American Advertising*. Minneapolis: Lerner, 2000.

Miles, Steven. *Consumerism: As a Way of Life*. Thousand Oaks, CA: SAGE, 1998.

Mintz, Steven, with Susan M. Kellogg. *Domestic Revolutions: A Social History of American Family Life*. New York: Free Press, 1989.

Pelfrey, William. *Billy, Alfred, and General Motors*, New York: AMACOM, 2006.

Pollack, Larry. *Why the PBGC Termination Insurance Program Should Be Ended*. Schaumburg, IL: Society of Actuaries, 2005.

Popkin, Jeremy D. *A History of Modern France*, 3rd ed. London: Routledge, 2003.

Porter, Glenn. *Raymond Loewy: Designs for a Consumer Culture*. Wilmington, DE: Hagley Museum and Library, 2002.

———. *Raymond Loewy's Designs in Motion*. Wilmington, DE: Hagley Museum and Library, 2003.

Pulos, Arthur J. *American Design Ethic: A History of Industrial Design*. Cambridge, MA: MIT Press, 1986.

———. *The American Design Adventure, 1940–1975*. Cambridge, MA: MIT Press, 1988.

Reed, Robert. *The Streamlined Era*. San Marino, CA: Golden West Books, 1975.

Schoenberger, Angela, ed. *Raymond Loewy: Pioneer of American Industrial Design*. Munich: Prestal Verlag, 1990.

Slater, Don. *Consumer, Culture and Modernity*. New York: Wiley, 1999.

Teague, Walter Dorwin. *Design This Day*. New York: Harcourt Brace, 1940.

Teich, Miklas, and Roy Porter, eds. *The Industrial Revolution in National Context: Europe and the USA*. Cambridge: Cambridge University Press, 1996.

Tretiak, Philippe. *Raymond Loewy and Streamlined Design*. London: Universe Publishing / Vendome Press, 1999.

Van Doren, Harold. *Industrial Design: A Practical Guide*. New York: McGraw-Hill, 1940.

Walsh, Kenneth T. *Air Force One: A History of the Presidents and Their Planes*. New York: Hyperion Books, 2004.

Weil, Gordon Lee. *Sears Roebuck USA: The Great American Catalog Company and How It Grew*. New York: Jove, 1977.

Wendel, C. H. *100 Years of International Harvester*. Minneapolis: Motorbooks International, 1993.

Worthy, James C. *Shaping an American Institution: Robert E. Wood and Sears Roebuck*. Champaign: University of Illinois Press, 1984.

Index